# FIRST CONTACT

## The Search for
## Extraterrestrial Intelligence

# FIRST CONTACT

## The Search for Extraterrestrial Intelligence

### Edited by
### Ben Bova and Byron Preiss

**Science Editor**
William R. Alschuler

**Associate Editor**
Howard Zimmerman

A Byron Preiss Book

NAL BOOKS

NAL BOOKS
Published by the Penguin Group
Penguin Books USA Inc., 375 Hudson Street,
New York, New York 10014, U.S.A.
Penguin Books Ltd., 27 Wrights Lane,
London W8 5TZ, England
Penguin Books Australia Ltd., Ringwood,
Victoria, Australia
Penguin Books Canada Ltd., 2801 John Street,
Markham, Ontario, Canada L3R 1B4
Penguin Books (N.Z.) Ltd., 182-190 Wairau Road,
Auckland 10, New Zealand

Penguin Books Ltd, Registered Offices:
Harmondsworth, Middlesex, England

*First published by NAL BOOKS, an imprint of Penguin Books USA Inc.*
*Published simultaneously in Canada.*

*First Printing, April, 1990*
*10 9 8 7 6 5 4 3 2 1*

**NAL BOOKS** REGISTERED TRADEMARK—MARCA REGISTRADA

ACKNOWLEDGMENTS:
The editors would like to gratefully acknowledge the following people for their help in putting this book together:
At New American Library—John Silbersack, Arnold Dolin, Gary Luke and Morris Taub. At Byron Preiss Visual Publications—Mary Higgins,
Patty Delahanty and David Keller. Also, thanks for their invaluable help to Jill Tarter at the SETI Institute, and Steve Maran at the press
office of the American Astronomical Association.

Library of Congress Cataloging-in-Publication Data

Printed in the United States of America
Designed by Steve Brenninkmeyer, Elizabeth Wen.
Set in Eurostyle, Garamond Book

"Mystery of the Great Silence" copyright 1990 David Brin, is based on and expanded from "Xenology: The New Science of Asking
Who's Out There" (Analog, 1983, copyright Davis Publications) and "How Dangerous is the Galaxy?" (Analog, 1985, copyright Davis
Publications).

A Martian Odyssey: Copyright 1934 by Continental Publications; copyright 1936 by Margaret Weinbaum; reprinted on behalf of the
Weinbaum Estate by authorization of its literary representative, Forrest J Ackerman, 2495 Glendower Ave., Hollywood, CA 90027.

"Where Are They?" is copyright 1986 by Arthur C. Clarke.

"Reflections" copyright 1990 by Philip Morrison is based on his essay for *Life in the Universe* (MIT Press, 1981).

"Answer, Please Answer" copyright 1990 Ben Bova, first appeared in *Amazing* magazine (Ziff-Davis, copyright 1962).

Opening each chapter is the footprint of Apollo 11 astronaut Neil A. Armstrong in the lunar dust, taken on July 20, 1969. *(Photo:
Courtesy NASA.)*

Title Page: The whole celestial sphere, as seen from Earth, in galactic coordinates. The Milky Way (center) shows the plane of our galaxy.
The tail of the constellation Scorpius and the teapot of Sagittarius are just above and below the galactic center. The small, slanting "dash"
below the Milky Way at left is M 31, the Andromeda Galaxy—the most distant object visible to the naked eye. *(Sky map: Lund
Observatory, Courtesy the American Museum of Natural History.)*

# CONTENTS

# PREFACE

## BY BYRON PREISS

This book was in gestation when a work entitled *Communion* shot up the American bestseller lists. As a child of the Fifties, I was astounded to see such a quaint story generate so much attention. After all, I had absorbed my share of "Mars Attacks!" bubblegum cards, B-movie aliens and UFO headlines. What made Whitley Streiber's account so special?

It was no accident, I deduced, that a few years earlier, *E.T.—The Extraterrestrial,* became the most successful film of all time. There is a persistent universal hunger to feel that we are not alone in the Galaxy, and *E.T.* fed that feeling in the most charming way, through the eyes of a child. *Communion* addressed that hunger in its own way.

What distressed me about *E.T.* and *Communion,* however, was that while they made people think about the subject of extraterrestrial intelligence, they did relatively little to bring attention to the real search for extraterrestrial intelligence— SETI—which was already in existence.

Since the Fifties, tremendous leaps forward have been made in our ability to seek out cosmic neighbors *and* to understand an alien intelligence, if we beat the odds and actually find it (or if it finds us).

SETI was no longer science fiction, it was science. More rewardingly, SETI was science in which we could all participate, whether through financial support or individual action. It was one of those times when reality exceeded fiction. *E.T.* was fun, but SETI was real.

Numerous excellent books had been written about SETI, but most were theoretical. I wanted to produce a book with an activist spirit, written by the men and women whose work had contributed to humanity's preparedness for first contact, and I was fortunate to convince Ben Bova to be my co-editor. As a former editor of

OMNI and president of the National Space Society, Ben has experience making yesterday's science fiction into today's public awareness of space science. He also has one of the most respected reputations in the field.

As our science editor, we recruited Dr. William Alschuler, president of Future Museums, whose diligence and academic credentials helped convince some of the more elusive figures of the SETI movement to contribute to the book. As associate editor, former *Future Life* magazine editor, Howard Zimmerman, ensured that the text would be comprehensible to the nonspecialist, and contributed significantly to the structure of the book. Gary Luke, our editor at New American Library, was a caring and supportive party through the editorial process. I would also like to thank David Keller, Patty Delahanty, Elizabeth Wen and Steve Brenninkmeyer for their contributions.

I hope this book will be a precedent to a new body of scientific literature: a hands-on approach to meeting our extraterrestrial counterparts. In the meantime, we fervently hope for the protection of all intelligent species on Earth, from the dolphin to the mountain gorilla, whose lives have been threatened by our own actions. Humanity's relationship with its fellow species is, after all, our most important first contact. For all we know, in their precious minds may exist some of the tools to understanding future visitors to our planet.

# FIRST CONTACT

## The Search for
## Extraterrestrial Intelligence

# INTRODUCTION

# SURVEYING THE COSMOS

## BY BEN BOVA

Look out at the sky on a moonless night. Not from the streets of a brightly lit, dusty, and polluted city, but from an open field far out in the countryside. Or from the chill dark of the desert, the kind of barren emptiness where astronomy began long millennia ago. Or, best of all, from the deck of a ship in the middle of the ocean.

There you can see the sky in all its splendor. Stars fill the darkness wherever you look, and the glowing sweep of the Milky Way arches overhead like a river of faintly shimmering light.

You will be able to see more than two thousand stars on such a night with your unaided eyes. Which of them might be the home of an intelligent civilization? Which ones should we direct our attention to as we search for the first contact with extraterrestrial intelligence?

In 1609 Galileo became the first astronomer to turn a telescope toward the stars. He found that there were far more than the few thousand we can see with the naked eye. Stars that were invisible to the unaided eye showed up even in his little thirty-power telescope. The Milky Way itself was revealed as a vast assemblage of stars, so many stars so distant from us that their light blends together to form the band of brightness that we see.

As astronomers built larger and larger telescopes they found more and more stars. Giant stars. Dwarf stars. Some stars were blue, some red, some yellow like our own Sun. Some pulsated as if they were breathing in and out.

## OUR GALAXY AND OTHERS

By the middle of the twentieth century astronomers had learned that the Milky Way is actually a vast, pinwheel-shaped *galaxy* that contains more than a hundred billion ($10^{11}$) stars and enough loose gas and dust to create billions more. And there are billions of other galaxies, far beyond our own, many of them much larger than our Milky Way.[1]

The distances to the stars in our galaxy are immense, mind-boggling. Even light, the fastest thing in the Universe, takes years to travel between the stars. Light moves at 186,000 miles per second (300,000 kilometers per second). It takes light eight minutes to go from the Sun to Earth, a distance that averages roughly 93 million miles. Light from the Sun reaches Pluto, the farthest known planet of our solar system, in five hours and thirty minutes.

To reach the nearest star, Alpha Centauri, light from the Sun must travel 4.3 *years*. If you drew a map in which the Earth-to-Sun distance was shrunk to one inch, the distance between the Sun and Alpha Centauri would be 4.3 miles. And that is the closest star to our solar system.

Astronomers use the *light-year* as a yardstick to measure distances in space. The distance that light travels in one year is close to 6 trillion miles (nearly 10 trillion kilometers).[2]

The Milky Way Galaxy is a vast double spiral of stars about 100,000 light-years across and 12,000 light-years thick at its core. Our solar system lies off to one side of the Milky Way, roughly 30,000 light-years from its center, where the Galaxy is about 2,000 light-years thick. The Sun travels around the galactic center in an orbit that takes 250 million years to complete.

With a hundred billion stars in our galaxy alone, which are the most likely to harbor intelligent life? What regions of the sky should we be

---

*1. Beware of Hollywood's use of the word* galaxy. *In most motion pictures the word is mistakenly used to mean a solar system, a system consisting of a star and its planets. The proper astronomical definition of the word* galaxy *is a system of billions of stars, such as our own Milky Way or the great spiral in Andromeda.*

*2. Astronomers also use a slightly larger unit, the* parsec, *in measuring distances. One parsec equals 3.26 light-years. Think of the light-year as the astronomers' "foot" and the parsec as their "yard."*

searching to find signals from alien civilizations? Or are we alone in the Galaxy, perhaps alone in the Universe? Could it be that intelligence, or even life itself, is a fluke, a statistical accident that is so unlikely it has occurred only once?

What do the stars tell us?

STELLAR LIFE CYCLES

The first and most impressive fact is that there are so *many* stars. It is difficult to believe that around all those billions of billions we are the only living creatures, the only intelligence.

Stars come in many sizes and colors. Some appear to be single stars, such as our Sun. Others are double. Sirius, the brightest star in our sky, is a double star. There are triple star systems (Alpha Centauri is a triple), and even more complex systems of multiple stars.

A star's color is a clue to its surface temperature. Astronomers have set up a classification system that ranks color and temperature. (See Table 1.) The system originally was strictly alphabetical, with A representing the hottest stars, and so forth. Over the years, though, it was found that B stars are actually hotter than A, so some classifications were dropped and others added until the current lineup evolved: O, B, A, F, G, K, M. Astronomers have no trouble remembering the order. They simply recall the mnemonic, "Oh, Be A Fine Girl, Kiss Me!" Obviously all this happened before the women's movement hit astronomy.

Astronomers have learned much about the life cycles of stars. Yes, stars have life cycles much as human beings do. Stars are born, go through a certain lifespan, then weaken and die. Some of them die in violent explosions. For most stars the life span is measured in billions of years. Our Sun, for example, is probably a bit less than five billion years old and has at least another five billion years of steady, dependable life ahead of it before it becomes unstable and ultimately collapses and fades out.

How can astronomers learn the life histories of stars that take billions of years to play out? We only see the stars in an instant of time; even several centuries of observation is little more than a snapshot. The very multiplicity of the stars comes to the astronomers' aid. By taking snapshots of thousands, millions, billions of stars, the astronomers can put them together into a sort of family album that, coupled with computer models of stellar evolution, shows young stars, old stars, middle-aged stars—stars in every stage of their eons-long life cycles.

## STAR BIRTH, STAR DEATH

Astronomical cameras have captured new stars in the midst of birth. *Protostars*, dark clumps of interstellar gas and dust, have been photo-graphed in the process of formation. A few years later the telescopes were turned to the same protostars and they were no longer dark! They had begun to shine. New stars had been born.

Stars begin as cold, dark clouds of gas and interstellar dust. Such clouds condense until their central regions become so hot and compressed that hydrogen in the core is forced to begin fusion reactions, creating helium and the energy that we ultimately see as starlight.

A star remains stable as long as it is fusing hydrogen into helium at its core. For most stars this stable middle age spans billions of years. However, the rate at which hydrogen is consumed depends on the mass of the stars. Hot blue giants such as Rigel and Spica are burning up their core hydrogen so fast that they will have stable life spans of only a few hundred million years, at best, instead of the billions of years that less massive, cooler stars have.

Stars also die. When they have used up the energy sources within them they collapse and go dim. Some of them explode so violently as they collapse that they can outshine an entire galaxy—briefly. Such *supernova* explosions are rare: Perhaps one occurs in our galaxy every five hundred years, on average. In 1987 a supernova blasted out in the Large Megellanic Cloud, a companion to the Milky Way, the closest supernova observed since 1604.

When a star dies its fate is determined by its original mass. The Sun will not explode, according to astrophysical theory. Some five billion years from now the Sun will have exhausted its core hydrogen and will begin to swell and turn into a red giant star. Its bloated outer envelope may engulf the Earth itself. Even if it does not, in its red giant phase our friendly Sun will glow so fiercely that life on Earth will become impossible.

Ultimately the Sun will collapse, going from a red giant to a white dwarf. It will shrink to about the size of the Earth, a hundreth of its present diameter. The matter inside the Sun will become so compressed that a spoonful of it would weigh thousands of tons. Then slowly, over countless eons of time, the Sun will cool down and go dark, a burned-out stellar cinder.

Many white dwarfs have been seen. The star Sirius is accompanied by a white dwarf, a star with a mass similar to the Sun's but a size of only a few thousand miles across. Since Sirius is known as the Dog Star, its tiny companion has been dubbed the Pup.

More massive stars undergo more spectacular death throes. A supernova explosion releases as much energy in twenty-four hours as the Sun does in a billion years! Most of the star's material is ejected into space, and any planets that might have been orbiting the star are undoubtedly demolished utterly. Yet death can lead to new life: The material ejected by the earliest supernovas enriched the interstellar medium with heavy elements that eventually became the building blocks for new stars.

Nor does a supernova explosion entirely destroy the star. For a star with an original mass of more than about twice the Sun's, the core collapses into matter so dense that a spoonful would weigh billions of tons. The atomic particles are squeezed so hard that they coalesce to form neutrons. The star shrinks to a mere few miles in diameter and is composed entirely of neutrons (with, perhaps, a surface skin of pure iron).

Neutron stars have been observed despite their diminutive size, thanks to the fact that many of them are *pulsars*, which emit powerful beams of radio energy in regularly timed bursts. The first pulsar, discovered in 1967, put out such precise pulses of radio waves that at first astronomers thought they might have discovered signals from an extraterrestrial civilization. The radio pulses, each ten to twenty milliseconds long, came every 1.33730113 seconds. Their timing was more accurate than the finest quartz watches.

Further study, however, showed that the pulsars are natural phenomena, not signals from an alien life form. The key piece of evidence proving that pulsars are neutron stars came from the Crab Nebula, the visible and expanding remains of a supernova that was observed by Chinese and Japanese astronomers in A.D. 1054. In 1969 astronomers found a pulsar at the heart of the Crab Nebula's wildly distorted cloud of gas. Not only does it emit bursts of radio energy, it even winks on and off visibly, as shown on a synchronized TV detector.

Stars of more than about 8 times the Sun's mass undergo an even stranger fate when they reach the end of their life cycles.

When the core of such a massive star collapses in a supernova explosion, more than 3 solar masses remain and not even the formation of neutrons can stabilize it. Most of the star's matter is ejected into space in the supernova explosion, but the core of the star shrinks under the immense force of gravity even beyond the tiny size of a neutron star. The core keeps on collapsing and disappears into a *black hole*. The gravitational force is so titanic that the star's core is pulled down into a dimensionless point.

Not even light waves can escape the incredible gravitational force. The star disappears from our Universe. As the old saw says, the star "digs a hole, jumps in, and then pulls the hole in after it."

Astronomers have detected pinpoints in the sky that are sources of fierce x-ray emanations. They deduce that these locations are sites of black holes, where some of the gases surrounding the exploded stars are being sucked into insatiable gravitational wells, generating x-rays as the stars fall into oblivion.

## PLANETS OF OTHER STARS

The one form of life that we know of exists on the surface of a planet. Therefore we should seek out other planets as possible abodes of extraterrestrial intelligence.

Spacecraft have investigated all the planets of our solar system except tiny, distant Pluto. Although the chances that some form of life may exist on Mars cannot be absolutely ruled out, and we simply do not yet know enough about the outer planets and their moons to determine if life exists there, it seems clear that no *intelligent* life exists on the other worlds of our solar system.

For extraterrestrial intelligence we must look to the stars.

The stars are so far away that any planets they may harbor are too dim for ground-based telescopes to detect. Astronomers on Alpha Centauri (if any exist there) could not see the Earth, nor even detect giant Jupiter, with the same kind of telescopes we have on Earth.

However, the InfraRed Astronomy Satellite (IRAS), launched in 1981, detected clouds of dust around several stars, including bright Vega. Astronomers conclude that such dust clouds are either the raw materials for building planets or the leftovers after the process of planet-building is completed. Interplanetary space in our own solar system is strewn with dust; we see some of it in the *zodiacal light*, a band of faint luminosity along the line of the Sun's path through the sky created by the reflection of sunlight off the dust particles. (You need a dark night, far from the city, to see it.)

The Hubble Space Telescope, a ninety-four-inch optical telescope orbited by NASA's space shuttle, may have the resolving power to pick out planets circling the nearest stars. (See Chapter 4.)

Until it does we will have to be satisfied with theory and inference. Astrophysical theory states that many stars probably form planetary systems. And observations of some of the nearer stars show that their paths are wobbling slightly as they move through space, weaving back and forth as though something invisible is tugging at them. (See Chapter 3.) The invisible objects may well be planets. Astronomers, trained to be conservative in their pronouncements, call the unseen objects *dark companions.*

Of the forty stars nearest the Sun, four have been suspected to have dark companions. Eleven others are double or triple stars, which makes it very difficult to determine whether dark companions are present or not.

Of the five closest stars, Alpha Centauri is a triple, Barnard's Star and Lalande 21185 may have dark companions, Sirius is a double star, and Wolf 359 appears to be alone.

One other piece of evidence may indicate that many stars are accompanied by planets.

The Sun spins on its axis very slowly, roughly once a month. Many stars spin hundreds of times faster. Astrophysicists have calculated that, although the Sun may have spun quite rapidly at first, its rotation was probably slowed by the processes that created the planets and the drag of the primordial solar wind. Fast-spinning stars, perhaps, either have no planets or are too young for planetary systems to have formed around them and slowed their spin. (See Chapter 3.)

There is an enormous number of slow-spinning stars in the Milky Way. As Table 2 shows, G-type stars such as our Sun tend to spin quite slowly. Has their spin rate been slowed by planetary systems that we cannot yet see?

## THE CONDITIONS FOR LIFE

Even if *all* the stars of the Milky Way were accompanied by planets, that does not mean that every planet would harbor life—or intelligence.

It took nearly five billion years for intelligence to arise on Earth. During all that time the Sun has remained steady in its output of energy, and the Earth has remained in a stable orbit around the Sun.

If the Sun's energy output had fluctuated significantly—if it had grown suddenly much brighter or dimmer, or exploded even mildly—life on Earth would have been wiped out. Our life depends on having liquid water available. Only a few-percent change in the Sun's energy output could freeze the oceans or boil them away.

And the Earth's orbit is very nearly circular. This means there is comparatively little change in global temperature during the course of a year. If our orbit took Earth as far from the Sun as Mars is, and in as close as Venus, our water would alternately freeze and boil. The chances for life would be slim.

The Sun has remained stable for five billion years and has at least another five billion years of placid stability ahead of it. Not all stars are so well behaved. Table 2 shows that the hottest stars are stable for only a relatively short time, perhaps too short for life to arise, certainly too short to expect intelligence to evolve—assuming Earth-like rates of evolution.

The very bright giant stars such as Rigel are not stable long enough for life to flower. Not until we consider the cooler main-sequence F and G class stars do we find stable life spans long enough for life to develop. (Remember, we are using the only example of life and intelligence that we know of—ourselves—as our yardstick here.)

Curiously, it is exactly these cooler main-sequence yellow, orange, and red stars that show the slow spin rates that may be associated with planetary systems. So on two counts we can reject the hot young giant stars as likely abodes for intelligent life.

## ROOM ENOUGH FOR LIFE

Life needs an abode that is "thermally habitable."

The temperature of a planet must be right for life to exist on it. Earth is thermally habitable for life based on liquid water. Jupiter may be also, deep beneath its outer cloud deck, where heat upwelling from the Jovian core may have created a vast planet-wide ocean of water. Jupiter's moon Europa appears to be covered with water ice; there is some reason to suspect that liquid water may exist beneath the frozen surface.

The farther planets of our solar system are apparently too cold for liquid water. They may be suitable for life based on ammonia or methane, however.

A star will heat up objects within a certain space around it, depending on the star's surface temperature. There is a zone around each star, then, that can be called thermally habitable for a given form of life.

For our own solar system, the thermally habitable zone for liquid-water life extends from somewhere between the orbits of Venus and Earth barely out to the orbit of Mars. Another thermally habitable zone for possible ammonia-based life includes Jupiter and Saturn, as well as their moons, and possibly Uranus and Neptune as well.

To be a good prospect for having a life-bearing planet, a star should have as large a thermally habitable zone (or zones) as possible. (See Chapter 1.) The larger the zone the better the chances of one or more planetary orbits being in it. Also, the planet's orbit should remain within the zone all the

Table 1   STAR CLASSES

| Class | Surface Temperature | Color | Examples |
|---|---|---|---|
| O | above 45,000°F | blue-violet | rare |
| B | 45,000–20,000 | blue | Rigel, Spica |
| A | 20,000–13,500 | blue to white | Sirius, Vega |
| F | 13,500–11,000 | white to yellow | Canopus, Procyon |
| G | 11,000–9000 | yellow | Sun, Capella, Alpha Centauri A |
| K | 9000–7000 | orange | Arcturus, Aldebaran, Alpha Centauri B |
| M | less than 7000 | red | Betelgeuse, Antares, Alpha Centauri C |

Table 2   LIFE SPANS AND SPIN RATES OF STARS

| Star Class | Stable Life Span | Typical Spin Rate |
|---|---|---|
| B | 8 to 400 million years | 60 to 90 miles per sec. |
| A | 400 million to 4 billion years | 30 to 60 miles per sec. |
| F | 4 to 10 billion years | 30 to 60 miles per sec. |
| G | 10 to 30 billion years | 0 to 30 miles per sec. |
| K | 30 to 70 billion years | 0 to 30 miles per sec. |

time; it is difficult to imagine life evolving on a planet that is thermally habitable only part of its year. Planets associated with multiple star systems may have complex orbits that swing them in and out of one or more thermally habitable zones.

The stars with the largest thermally habitable zones are obviously the hot, young giants. But these stars are short-lived. They do not remain stable long enough to allow life to develop. The dinosaurs never saw Rigel or Spica; they were not shining yet. Such bright blue giants will be long gone before any planets near them have had the time to evolve life.

The stars with the longest stable lifetimes are the cool, dim red dwarfs. They burn their hydrogen fuel so slowly that they can remain stable for tens of billions of years. But they are so cool that their thermally habitable zones must be small, and the chances of having a planet orbiting inside such a zone seem slim.

It is no surprise, then, that we live on a planet that orbits a yellow star. Such stars are warm enough to have a respectably sized thermally habitable zone, and long-lived enough to allow life—even intelligence—to evolve.

There are billions of such stars in the Milky Way Galaxy. Of the forty stars closest to us, there is one F type, three G types, and twenty-nine K and M stars. This count includes only the brightest members of double and multiple star groups. Four of these stars are suspected of having unseen dark companions. Each of them might have Earth-type planets orbiting within thermally habitable zones, bearing life and even intelligent civilizations.

THE GEOGRAPHY OF THE MILKY WAY

As a starting rule of thumb, lacking evidence except for the example of ourselves, we can say that we should not expect to find intelligence on planets circling stars that are less than roughly five billion years old.

We know that the young giant stars do not meet this criterion. The cool K- and M-class dwarf stars have much longer life *expectancies* than the Sun. But are they actually older than the Sun? We know that some are not; some of them are just becoming stable after their formation out of protostar globules. Moreover, the thermally habitable zones of K and M dwarfs must be perilously slim.

Looking further afield, let us consider the "geography" of the Milky Way Galaxy in our quest to determine where extraterrestrial intelligence may be found.

The Milky Way is a spiral galaxy very much like the beautiful spiral in

Andromeda, M31.[3] The most distant object visible to the naked eye, M31 is about two million light-years away.

The core of our own Milky Way Galaxy (which lies in the direction of the constellation Sagittarius) must be thick with stars, as the core of M31 and other spiral galaxies are. But we never see the core optically because it is hidden behind thick clouds of interstellar dust. Radio and infrared observations have been able to penetrate the dust to some extent. These observations, plus the studies of the cores of other spiral galaxies, show that the region is so rich with stars that they are probably no more than a single light-year apart, at most.

Such studies of the Milky Way's core also indicate that something very energetic is blazing away at the heart of our galaxy. Possibly a giant black hole is gnawing away at the Milky Way's core, gobbling whole stars while discharging radio, infrared, and high-energy radiation.

Most of the stars at the cores of spiral galaxies are much older than the stars in their arms. Red giant stars are common in the core regions, and astrophysical theory shows that stars become red giants only after they have used up most of their original hydrogen and left their long life of stability behind them. There are few hot young blue giants in galactic cores; these are found almost exclusively in the spiral arms.

Because the core regions of spiral galaxies seem to be populated predominantly by different types of stars than the spiral arms, astronomers refer to the two different stellar constituencies as Population I and Population II. This can cause some confusion.

Population I stars are the kind found in the spiral arms. Our Sun is a Population I star. These are youngish stars; their brightest members are blue giants such as Rigel. Population I stars contain a relatively high proportion of elements heavier than hydrogen and helium, although "relatively high" never amounts to more than a few percent. Yet astronomers refer to the Population I stars as "metal rich." ("Metal," in this usage, means any element heavier than helium.)

---

3. *M31 is the thirty-first item of 107 objects listed in the 1784 catalogue of Charles Messier, a French astronomer who especially sought to discover new comets. His catalogue was originally intended as a guide to fellow comet-seekers, identifying fuzzy objects that are not comets and not worth bothering with. Or so he thought. Thirty-four of the "Messier objects" are galaxies, the rest star clusters or interstellar clouds within the Milky Way.*

Population II stars are those found in the core regions of a galaxy. They are old. Their brightest members are red giants, stars that have gone through many billions of years of stability and have now become bloated and swollen. Population II stars are mostly "metal poor."

The heavy-element content of a star is an important clue to its history. Why are the stars in the Milky Way's core metal-poor while the stars in the spiral arms are metal-rich? Because the elements heavier than helium have been created inside the stars. Consider the following:

The Milky Way must have been quite different ten or fifteen billion years ago—a span of time two to three times longer than the Sun's age. It is almost misleading to use the name Milky Way that far back in time, because there were no stars to make it shine.

Our galaxy presumably began as an immense dark cloud of gas, at least 100,000 light-years across. The gas was almost entirely hydrogen, the lightest and simplest of all the elements, probably with a smattering of the second-lightest element, helium.

## CREATION OF THE CHEMICAL ELEMENTS

The first stars to form, then, had no significant amounts of elements heavier than helium. All the heavier elements, up to iron, were "cooked" inside the stars. During the long eons of a stable star's life its fusion processes transmute hydrogen into helium. When the core hydrogen supply becomes depleted the star begins "burning" helium to create carbon, oxygen, and neon. These elements eventually are themselves used as fuel for further fusion processes that create still-heavier elements.

Once helium-burning begins at a star's core its outer envelope begins to swell. The star grows into a red giant.

When the star has reached the point where its fusion processes are creating atoms of iron, it has also reached the end of its tether. When iron undergoes fusion the process *absorbs* energy rather than releasing it. The star is suddenly bankrupt. It collapses, either into a white dwarf (as the Sun eventually will) or into the explosive fury of a supernova, as discussed above. In the star-shattering cataclysm of a supernova, the elements heavier than iron are created.

The earliest stars in the Milky Way began with nothing more than hydrogen and helium. They created the heavier elements and spewed them into interstellar space when they exploded. These atoms served as the building blocks for the next generation of stars. The explosions that marked

# THERMALLY HABITABLE ZONES

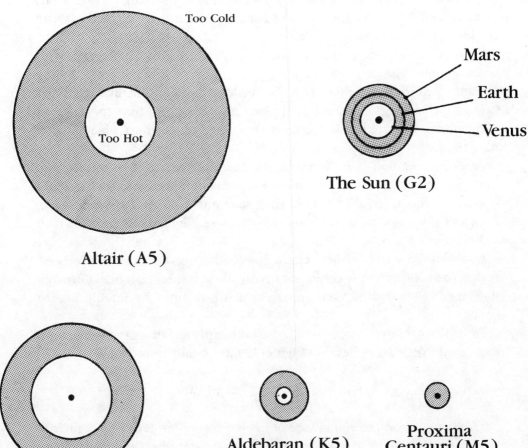

**Figure 1.** The shaded region around each star represents the orbital area in which temperatures are moderate enough that life may be possible. The orbits are all drawn to the same scale. *(Diagram supplied by Ben Bova. Art by Elizabeth Wen.)*

the deaths of the earliest stars enriched the interstellar medium with clouds of heavy elements. It was from those clouds that new stars were created.

Judging by the heavy-element content of the stars, most astrophysicists estimate that the Sun must be at least a third-generation star, a grandchild of the original stars of the Milky Way. The elements inside the Sun were created in the cores (and supernova explosions) of ancient stars. The atoms that make up the solar system were once thousands of light-years away, deep inside other stars. The atoms of our own bodies were created in those distant stars. We are truly stardust.

Those earliest stars, however, *could not produce life*. They began with nothing but hydrogen and helium. If there were any planets orbiting around them, those planets could be made of nothing but hydrogen and helium. Nothing else existed! There was no carbon, no oxygen—none of the elements that are needed to create the complex chemical phenomenon we call life.

Second-generation stars? It is possible that they would have most of the heavier elements, including the nitrogen, potassium, iron, and such that are needed for life to exist. Planets of such stars might be able to support life, even our own kind of water-based life, if those elements were present in sufficient quantities.

If life has arisen on planets circling second-generation stars there is no reason to suppose that it could not eventually attain intelligence. Certainly the long-lived red dwarf stars provide plenty of time for intelligence to develop.

Granting that intelligence could arise on the planets of second-generation stars, could such races develop high civilizations and technology?

## THE IMPORTANCE OF METALS

This depends on the abundance of natural resources. Fossil fuels such as coal and oil should be available, since these are the result of biodegrading plant and animal matter. What about metals? (And here we mean metals in the usual definition of the word, not the astrophysicist's.) Our technology here on Earth is built on copper, iron, nickel, tin, etc. Our history rings with the sounds of the Bronze Age, the Iron Age, the Steel Age, the Uranium Age.

Astronomical evidence is hazy here. Theory shows that second-generation stars should have a lesser abundance of metals than we third-generation types. But certainly there should be *some* metals on second-generation worlds.

How much metal is enough? There is no way for us to tell. Planets of second-generation stars might have iron mountains and gold nuggets lying on the open ground. Or they might have very little available metal. The planet Jupiter in our own solar system might easily contain more iron than the Earth does, but it would be mixed with 317.4 Earth masses of hydrogen, helium, methane, ammonia, and other elements, and buried at the bottom of an ocean that is thousands of miles deep. Not easily accessible.

If there are planets of second-generation stars where heavy metals are abundant and available, those planets could be sites for highly advanced civilizations. But what if intelligence arises on a world where heavy metals are not available?

First recognize that while intelligence per se does not depend on the existence of metals, *life* does. There is an atom of iron at the heart of every hemoglobin molecule in your bloodstream. Good health depends on trace amounts of iron, zinc, and other metals.

The human race rose to intelligence before the discovery of metalworking. Humans used bone, rock, clay, and wood for their earliest technology. Yet history shows that once metals became available our race took giant strides forward. Metals allowed our ancestors to build effective plows. And swords. And chariots. And radio telescopes. Today our skyscrapers and computers and engines and spacecraft are made largely of metals. Metals are strong, tough, and cheap.

Could an intelligent race build spacecraft and computers without metals? Could they build radio telescopes with which to receive our signals or send their own?

Today we humans use Space Age materials such as plastics and composites. But the machinery that produces them is made of stainless steel, copper, brass, and other metals. Cavemen, or even ancient Greeks, could not have produced boron-filament composites or polystyrene plastics. They did not have the metals with which to produce them.

Would a metal-poor second-generation intelligent race be stymied in its attempts at technology? Who can say? *Our* technology certainly depends on metals, and that is the only example we have to consider—so far.

Another important point: The entire world of electricity and magnetism would never have been discovered without copper and iron. It is difficult to see how the entire chain of study and application of the electromagnetic force—from Volta through Faraday, Maxwell, Hertz, and Marconi to radio telescopes and television and superconducting magnets—could have taken place on a metal-poor planet. And where would our technology be without electricity? Back in the mid-nineteenth century, at best.

It just might be possible for an intelligent race on a metal-poor world to build a complex technology out of nonmetals. However, tribes on Earth that never had easy access to metals never developed a complex technology. Coincidence, or cause and effect?

If our own history is any guide, it is the heavy metals that lead to high technology. They also form a natural gateway into the world of electromagnetism and the whole concept of "invisible" forces that act over a distance, such as gravity. We can trace a direct line from humankind's use of heavy metals to electromagnetics, nuclear power, space exploration, and the search for extraterrestrial intelligence.

Indeed, as we scan the stars with our radio telescopes, we are implicitly seeking intelligent races that have developed a technological civilization, a civilization that can build radio telescopes, a civilization that uses metals.

If there are second-generation worlds rich enough in metals to give rise to technological civilizations, then their races must be much older and presumably wiser than we are. Second-generation races that do not have metals may be gamboling innocently through some local variation of Eden. They will not have radio telescopes to either receive or transmit intelligent signals.

First-generation stars, if any still exist, will have no life at all.

Add to this the fact that the core of the Milky Way is being bathed with lethal amounts of hard radiation from the black hole (or whatever it is) that is blazing away there, and we must come to the conclusion that the place to search for intelligent signals is in our own neighborhood, here among the spiral arms of the Galaxy.

Gone is the old science-fiction dream of an immense galactic empire, presided over by the older and wiser races that exist at the heart of the Galaxy. Perhaps there was one there once, a second-generation race that had enough metal available to expand outward. Perhaps, as suggested by writer/editor Stanley Schmidt, the cataclysm that is destroying the core of the Milky Way is the result of an industrial accident, some bit of interstellar technology that went awry. If so, the accident certainly wiped out the home worlds of the race that caused it.

Our search for intelligent life should begin, then, in our own region of the Milky Way's spiral arms. We should eliminate stars that are too young to have developed intelligence, and not spend too much effort on those that are too dim to have a wide thermally habitable zone.

That leaves stars close in physical makeup to our Sun: yellow or orange, late F-, G-, and K-type stars that are still living out their long eons of stability before they begin to swell into red giants.

That simplifies our search. It eliminates literally billions of stars.

And still leaves a few billion to examine.

# CHAPTER 1

# LIFE IN THE UNIVERSE

*Is there intelligent life on Earth? Before we begin to seek extraterrestrial intelligence we must arrive at some understanding of just what we mean by intelligence. Might there be intelligent non-human life on Earth?*

*Isaac Asimov surveys the question of intelligence in general and produces a soundly workable definition of just what we should look for when we turn our probing senses toward the heavens. Diana Reiss reports on the "alien" intelligence that lives here on Earth: the dolphins. Hal Clement gives a detailed prescription for what kind of worlds might harbor intelligent life.*

*The one fact that overwhelms all others is that we have only a single example of intelligent life on which to base all our conjectures: ourselves. While Copernicus removed the Earth from the center of the Universe and Darwin showed that humans are not separate from the other life forms on our planet, when it comes to intelligent life we are alone on the center of the stage.*

*So far.*

# TERRESTRIAL INTELLIGENCE

## BY ISAAC ASIMOV

Intelligence on Earth is not really rare. Octopi are quite intelligent if compared to other invertebrates; crows are quite intelligent if compared to other birds; primates are quite intelligent if compared to other mammals.

However, there is only one type of organism in existence that is sufficiently intelligent to have developed a complex technology, and there we are speaking of ourselves and our immediate ancestors, the various hominids.

It worked this way. The earliest australopithecines, about five million years ago, were no taller than chimpanzees, slighter in build, and with a brain no larger. The australopithecines had, however, developed a backward bend in the lumbar region of the spine that made it possible for them to walk erect easily and in preference to any other form of locomotion. That distinguished them from all four-legged animals. Bears and chimpanzees might walk on their hind legs temporarily, for instance, but not continually and not by preference.

Erect posture freed hominid forelimbs for holding and manipulating. By bringing objects to close examination by sense organs, and by manipulating those objects, hominids flooded their brains with information. Any chance mutation that increased the size or complexity of the brain and enabled it to process the information more efficiently had survival value and was selected for. The brain, therefore, evolved explosively, and human beings now have far larger brains than any ape.

What counts is not only a large brain but a comparatively small body. Elephants and whales have brains that are larger than human brains, but the bodies of those creatures are so massive that the brain/body ratio is small.

The human brain makes up 2 percent of the mass of the human body. Those of elephants and whales make up far less than 2 percent.

Some of the smaller monkeys (marmosets, for instance) have brain/body ratios actually larger than human beings, but the brain, in absolute size, is much smaller than ours.

Only dolphins have brains as large and complex as those of human beings (or somewhat more so, actually) and bodies as small. They might be of human intelligence, but they live in the sea so that their bodies are streamlined and they lack any appendages that are in any way equivalent to human hands.

The large-brained elephant does have an appendage, its trunk, that might almost be compared to a hand, but actually, two such appendages, not one, are needed for the development of technology-based intelligence. One appendage is needed for holding and one for manipulating. In the case of 90 percent of human beings, the left hand is preferentially used for holding and the right for manipulating. (It is reversed in the other 10 percent.)

The result is that *Homo habilis*, the first organism sufficiently human to be included in the genus "Homo," once it evolved nearly two million years ago, was capable of shaping stone into tools. Other animals might use tools, and some might even be able to modify natural objects to make them more fit for use, but only members of genus *Homo* have ever been able to deliberately modify something as ordinarily intractable as stone. It is with the development of stone tools that hominids first differentiated themselves from all other forms of life.

Then, about 500,000 years ago, probably earlier, *Homo erectus* learned to make use of fire. No non-hominid has ever, under any circumstances, learned to do anything with fire but flee from it. This matter of fire is an even sharper differentiation than stone tools. All technology depends on fire; depends, that is, on the use of sources of energy other than that in living muscle, sources that are dependable, portable, and human-made in any amount. Thus, wind and running water are examples of inanimate energy that can be used by any organism, but they are not dependable or portable and must be taken only as they are found by any organism without a technology. (Fire is not possible in water, which means that dolphins, however intelligent, cannot develop a technology.)

Now, then, how do we define intelligence? I see no point in trying to define what it *is*. Not only would that be too complicated a task, but it is unnecessary. Far better to define what it *does*.

For our purposes, a species is intelligent if it can develop a complex technology.

One usefulness of this definition is that it makes it unnecessary to delve into psychology and philosophy. One doesn't have to be concerned with inner being and inner thoughts. One merely looks at what is being accomplished.

Another usefulness of the definition is that we don't have to bother with intelligence on the individual, or even the minor group level. Undoubtedly, some human individuals are more intelligent than others, but which are more intelligent and to what degree depends upon the precise definition you care to give to individual intelligence and the manner in which you decide to measure it. There is enough uncertainty and controversy over both definition and measurement to give us a feeling of relief at not having to deal with it.

The fact is that an individual human being, or even a small group, left to itself cannot develop a technology, except very slowly as the group expands. (There are romantic stories, such as *Robinson Crusoe*, which seem to demonstrate the reverse, but Crusoe started with a shipload of equipment and the memories of a technology.)

Let us restrict intelligence, then, to nothing smaller in scale than the *species*, and to *the development of a technology* as the measure of intelligence for our purposes.

But what do we mean by a technology? Spiders build webs. Ants, bees, and termites build complicated dwelling places and develop complicated societies with specialized activities. Some social insects do things that look very much like agriculture, herding, enslavement, and so on. Beavers build dams and significantly change the environment in their own favor.

On the other hand, Paleolithic humanity had only the sketchiest kind of technology, nothing that we would consider an *advanced* technology. (And what is an *advanced* technology, by the way? What we have naturally seems advanced to ourselves, but another species with a far more advanced technology, or even we ourselves a thousand years from now, might consider our present technology primitive, and therefore useless as an indication of intelligence.)

But please note that I didn't define intelligence as the *possession* of a technology, but the *development* of a technology. In estimating whether a species is intelligent or not, it is insufficient to look at it and judge it at a specific time.

In every species of organism we know, other than the human being and its immediate ancestors, significant alterations in behavior take place only as a result of evolutionary change, which is very, very slow. A spider, building a web today, does so exactly as members of its species, or closely allied species, would have done a million years ago, or will do a million years from now.

Undoubtedly, web-building evolved and grew more efficient with time, but only at the general rate of biological evolution. The same is true for the most complicated of termite societies, for dam-building among beavers, and so on.

Even creatures as intelligent as gorillas, elephants, or dolphins live now, we suspect, as they have always lived, and will always live as long as they do not evolve. (Some monkeys have been reported to learn new varieties of behavior, which shows they have the beginnings of intelligence.)

Only human beings, *without* significant biological evolutionary change, can alter their behavior radically as a result of *cultural* evolution, through the accumulation of information, and the deliberate conception of "improvement."

Cro-Magnon man, twenty-five thousand years ago, had a brain as good as ours and was as intelligent as we. The measure of that intelligence is not the technology that the Cro-Magnons had or that we have, but the *change* in technology that has taken place in those twenty-five thousand years without any significant change in the quality of the brain. The change is slow at first, but it is cumulative and eventually becomes rapid; but even slow change is indicative of intelligence.

Zero change with time, whatever the appearance of technology, is what we might refer to as "instinct" rather than intelligence.

But if intelligence is to be found only in our own species and its immediate ancestors among all the tens of millions of species that have inhabited our world in the last three and a half billion years, is it possible that human beings can now put their intelligence to work to *create* an intelligent species?

We have built computers and sometimes these are considered "thinking machines" and people talk of the possibility of "artificial intelligence."

To begin with, what is a computer?

A computer is a device capable of doing things that, throughout history, we have associated with intelligence, and not with the instinct of animals or the mechanical behavior of machines.

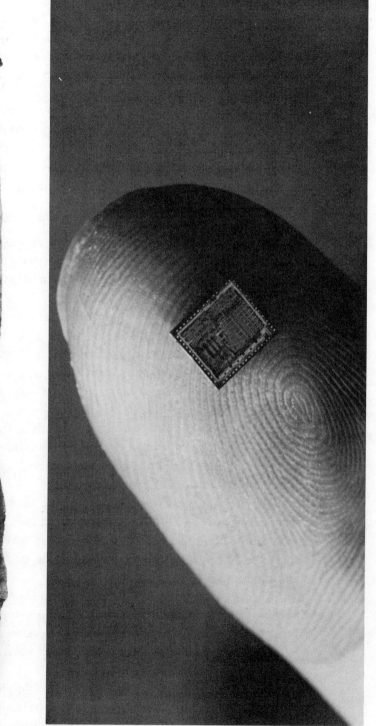

**Figure 1.** Chipped tools from the Paleolithic, and a modern computer chip. The continued development of a technology is the hallmark of intelligence. *(Courtesy American Museum of Natural History and Delco Electronics.)*

For instance, we associate the solution of mathematical problems with human intelligence. No other living organism can do it, and, until the 1940s, calculating devices have been simple indeed and have clearly and obviously worked only by something we might call "artificial instinct" rather than "artificial intelligence."

Modern computers, however, solve problems that are truly complex, and do so with such speed that human beings cannot compete with them at all. The temptation is to think of machines that can do *mental* labors that are far beyond us as certainly intelligent and even superintelligent.

And yet such computers show no true intelligence. A modern computer, however advanced, does nothing it is not directed to do. It is working with electrical impulses that do not get tired, do not make mistakes, and are transmitted at the speed of light. It is engineering that gives the *illusion* of intelligence. The human brain does get tired, does make mistakes, and works much more slowly, but that is the penalty of not being too-simply "wired," of not being driven by instinct.

A typical modern computer, set to working, will do what it does forever without change. Even if it is so designed as to improve its responses by considering past events, that improvement is sharply limited. No computer or group of computers existing now or in the foreseeable future can start from scratch and develop a technology, so they are not intelligent by the definition I have presented.

But computers do change, if not as a result of their own inner capacities, then because human beings are forever building new ones of improved design. Will we ever be the agents for the evolution of computers that *are* examples of true "artificial intelligence"?

I doubt it. One must first understand the true complexity of the human brain as it has evolved over three and a half billion years. The human brain consists of 10 billion neurons and 90 billion auxiliary cells. No computer, either now or in the foreseeable future, is going to contain 100 billion switching units.

And even if a computer were to contain so many units, the neurons of the brain are interconnected with extraordinary complexity, each being connected to dozens or thousands of others in a manner that passes our understanding. Computers don't have even the beginnings of such complexity.

And even if we learn to duplicate the complexity, too, then the fact remains that the units in computers are switches that move from on to off and back to on, and nothing more. The neurons of the brain, on the other

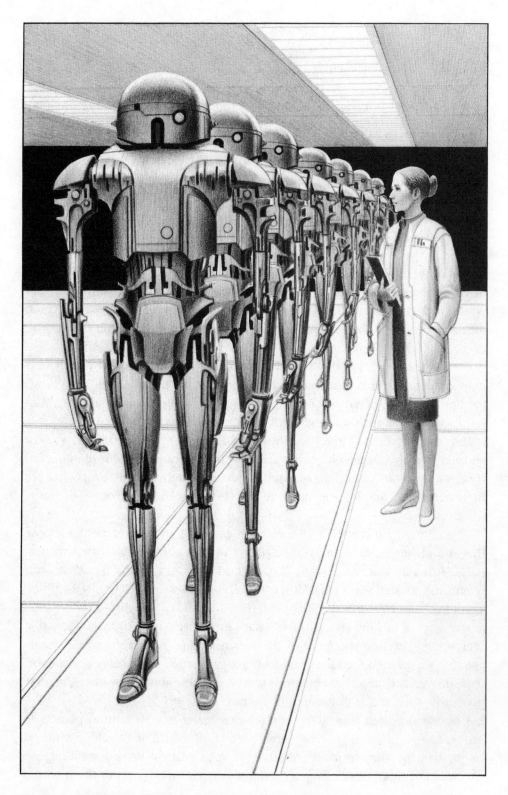

**Figure 2.** Intelligent, mobile robots play a big role in science fiction, such as these positronically designed characters from an early Asimov robot story. *(Art: Ralph McQuarrie.)*

hand, are enormously complex structures of macromolecules of various types whose functions we do not entirely understand.

True artificial *intelligence* is not something we can easily attain, or may even want to. It is much easier to produce more human beings (indeed, a major problem of humanity today is that we produce too *many* human beings and that there is a perilous oversupply) than to produce and multiply enormously complex computers.

Better to develop computers that are better and better at *artificial instinct*, leaving the intelligence to ourselves. Computers are tools and can become magnificent tools, and why should we want more than that?

It is in the highest degree unlikely, of course, that Earth is the only world in the Universe that has developed life. It seems quite possible than any world with the physical and chemical characteristics of Earth may develop life, and there may conceivably be millions of such Earth-like planets in each of billions of galaxies.

Even if this is so, we don't know where any of these life-bearing planets exist, and have no reasonable system (other than hit-and-miss searching) for finding them. (That has been called feeling about in a large dark room for a small black cat that might not be there in the first place.)

But then, even if there are life-bearing planets and, in one way or another, we encounter one, what are the chances of intelligent life upon it? If we cling to our definition of intelligence, we might ask instead: What are the chances of our finding upon it a technology in the process of being developed?

The only way we have a chance of guessing at an answer is to suppose that the planet Earth is an average planet in every respect and that we can judge from its one case what the general situation might be. Thus, our planet has existed for a total of 4,600 million years and it is only 2 million years ago that *Homo habilis* appeared and began to develop a technology at the level of making stone tools. What's more, it is only 200 years ago that technology reached the level of the steam engine. And it is only 20 years ago that we reached the Moon and technology reached the level of spaceflight.

If this is the kind of situation that is representative of the Universe generally, then technologies might be rare indeed.

I once calculated that there might be as many as 640 million planets in our galaxy capable of bearing life. Even if all of them developed life, it might well be that on only 280,000 of such planets would we find any species capable of developing a technology; that on only 28 of them would

we find an industrialized world; and that only 2 or 3 would have advanced
to the level of spaceflight. And the chances that any of them would have
advanced to the point where interstellar flight was practical would be
virtually zero.

But that supposes that all life-bearing worlds are as old as Earth and no
more. The Universe, however, was already 10 billion years when Earth was
formed. It might, by then, have formed many millions of life-bearing worlds
and, even at the most pessimistic, many dozens of technologies capable of
space flight even before Earth was anything more than part of the swirling
cloud of dust and gas that eventually formed our solar system. In that case,
there must surely have been time for alien intelligences to have explored
the Galaxy thoroughly.

Enrico Fermi, faced with that possibility, asked, "Then where is everybody?"

The fact that in Earth's history no aliens have reached us (as far as we
know, and never mind the UFO-maniacs) would indicate that other advanced technologies like our own are indeed excessively rare. Another
possibility is that whether or not they are excessively rare the problem of
interstellar flight defeats them all. Still another possibility is that we have
indeed been discovered but that the discoverers refrain from interfering
with us, knowing that technological societies are rare indeed and any that
might develop must be protected and cared for.

But suppose *we* are the ones who go out and explore and come across
life-bearing planets, with or without technologies? (Using my definition of
intelligence, we don't have to worry about having to decide whether some
alien species is intelligent or not, and fearing that its intelligence might be
so different from our own in kind as to be unrecognizable. We are looking
for *technology* and that, being a material system, should be simpler to
recognize than something as inchoate and as hard to pin down as abstract
intelligence.)

If technologies are rare or nonexistent, can we feel all the safer in
exploring the Universe? After all, what can nontechnological life do against
our sophisticated weapons? Well, ask yourself, how technologically advanced is the AIDS virus?

Any life-bearing planet might have its dangers. Possibly the least dangerous
are those that carry advanced technologies. Species that possess such technologies might well be fascinated by us and be much more anxious to learn
about us than to do us harm. And we, of course, could learn from them.

First contact between two advanced technological civilizations might
well result in a fusion of knowledge that might benefit each one extremely.

But what if we encounter a technology as far beyond ours as ours is beyond that of a chimpanzee? Does that mean the superior aliens would instantly wipe us out as despicable vermin? Is that how reasonable human beings feel toward chimpanzees? We have reached the stage where we are anxious to protect and preserve chimpanzees. We study them and try to teach them to communicate by gesture and are delighted with every small success. Why may not those more advanced than ourselves be at least as humane?

Will the very ability of the superior beings to care for us and condescend to us prove unbearable to us? After all, a chimpanzee probably isn't bright enough to appreciate how much brighter a human being is. We, however, are quite likely to be bright enough to realize our own inferiority, and that realization might destroy us.

Still, there is a difference between a more advanced technology, and a more advanced intelligence. A more advanced technology, however far beyond us, can be learned—just as Archimedes, if he were brought into our world as a young man, could easily learn our advanced mathematics.

And yet, what are the chances of encountering a more advanced intelligence? How intelligent can an intelligent species become?

We can easily imagine a larger brain, a more complex brain, a more efficient brain, one more capable of making leaps of creation, of grasping conclusions, of foreseeing consequences—of doing anything that we associate with high intelligence.

Yet as a brain grows larger and more complex, might it not also become more unstable, more capable of sudden and disastrous breakdowns? For all we know, the complexity of the brain and the intensity of intelligence may reach a point where the possibilities of breakdown match the benefits of creation, and after that further advance is impractical.

And—conceivably—we may ourselves be near that point at this time, so that the chances of meeting any species more intelligent is small indeed.

Maybe.

# THE DOLPHIN: AN ALIEN INTELLIGENCE

## BY DIANA REISS

In searching for evidence of intelligence in other species on this planet we often fall into the search by looking for intelligence as we know it, measuring intelligence by how closely other life forms do what we do. But surely other organisms living in other environments have to deal with different demands and types of information. We may have to broaden our definition.

The viewpoint that man is alone as an intelligent species is not only heard in discussions regarding our relationship to the rest of the biological world. Echoes of these same arguments are now being heard in discussions speculating about the probability of extraterrestrial intelligence. Theoretical discussions about extraterrestrial life have revealed a wide range of assumptions about intelligence. Many scientists have speculated that the rise of intelligence may be a convergent process, a widespread phenomenon throughout the universe. Others argue that "true" intelligence would only evolve in a handed humanoid species capable of developing language, tool use, and technology.

Bioastronomers, biologists, and psychologists face common problems in exploring non-human intelligence and communication with other species. These problems include: (1) signal detection and pattern recognition; (2) how we decode an attempt to communicate with other life forms; and (3) how we look for or recognize "intelligence" in non-human life forms.

Communication is a ubiquitous phenomenon found throughout the animal world. Complexity and plasticity (the ability to alter behavior according to context and as a result of past experience) are seen in the communication, behavior, and navigation systems of many animal species.

Field and laboratory studies have provided evidence that bees, birds, non-human primates, and dolphins either use or can learn to use symbolic or referential signals (i.e., signals that refer to other things) for exchanging information.

In the 1960s the public's imagination soared with reports about the high intelligence of dolphins and the possibility that communication could be established with these large-brained and highly social mammals. Much of this fascination was due to John Lilly's early research investigating the neurophysiology of the bottlenose dolphin (Tursiops truncatus), which provided information as to the structural complexity of their large neocortex (the most recently evolved layer of cortex, the outer layer of the brain thought to be involved in the integrative and associative processing of information). Lilly reported on the dolphins' ability for vocal mimicry and attempted to teach them to mimic English in his flooded laboratory. Lilly felt certain that in the years to come the barrier between human and dolphin communication would be broken. It is now thirty years later, and there are still no English-speaking dolphins.

While Lilly's later work became more speculative than scientific, it stimulated other scientists to investigate dolphin abilities from the perspectives of social ecology, behavior, signal processing, and communication. We are still intrigued by these highly social mammals and have quite a bit more information about their form of intelligence.

## THE DOLPHIN

The dolphin is a superb model for helping us formulate ways of describing and understanding intelligence in a non-human species. Dolphins and whales, members of the order of mammals called Cetacea, offer a unique opportunity to gain insights into evolutionary strategies in and adaptations to an environment in which we do not dwell.

Dolphins and other cetacean species differ from us in that they have adapted to their environment rather than altering their environment to suit their needs. The fossil record and skeletal features show the dolphin's ancestors were once terrestrial mammals, possibly evolved from a family of archaic ungulates, Condylarthra, related to cows. The dolphin made a return to an aquatic existence in the early Eocene period nearly sixty million years ago, just after the dinosaurs disappeared. The extent of its adaptations to this environment is remarkable. For example, skulls of archaic whales document a progressive migration of the nostrils from the rostral region (near the front of the head), typical in terrestrial mammals, to a blowhole at the top of the head, which facilitates surface breathing. Additionally, most of the typical mammalian appendages, such as ear flaps, tail, legs, erectile hairs, etc., that normally can be used to convey information between animals have been lost in the radical streamlining of the dolphin's body in adapting to an aquatic existence.

Then how do dolphins communicate? It is well known that dolphins use a wide variety of signals in their social interactions, including visual, tactile, acoustic, and perhaps gustatory channels. Captive dolphins exhibit a large repertoire of complex vocalizations categorized into three classes of signals: broadband "clicks" called sonar or echolocation, used for navigation and perception; other typically mammalian sounds such as squawks or barks; and narrow-band variable-pitch whistles believed to be used solely for social communication. To date however, we understand very little of how dolphins communicate, since the combination of body posture, head and body movements, tactile interaction, and acoustic signaling is a complex system. What we do know is that every dolphin has a personalized signature whistle used for individual recognition, and perhaps other functions as well. The apparent mimicry of signature whistles between dolphins has been reported recently in the scientific literature. In a field study of free-ranging bottlenose dolphins and in our research program investigating dolphin communication and cognition in our laboratory at Marine World, we have found that there is a similarity in the signature whistle of young male dolphins and their mothers. The function of this is not clear, but may be related to kinship ties.

In using echolocation, or biological sonar, dolphins emit click sequences of sonic (within the range of human hearing) and ultrasonic frequencies and rapidly process the reflected echoes to obtain "acoustic images" and additional information about their environment. It is believed that echolocation is used for navigation, exploration, foraging, and the detection and inspection of objects in the environment. Experiments have demonstrated that through sonar use dolphins can perceive the size, composition, position, and movements of objects. Some investigators have speculated that they may also use their sonar for obtaining information about the internal body state of other dolphins and other organisms. This naturally evolved sensory system surpasses our artificial sonar and radar technology. The non-handed dolphin has not developed a technology like ours that is reflected in our scientific and industrial achievements. But, if we define technology more broadly from an anthropological perspective as the body of knowledge available to a group that is of use in practicing arts and skills and extracting or collecting materials, then we have more of a basis for comparison.

Studies of brain physiology and anatomy have revealed that the dolphin brain is relatively large and highly convoluted. The dolphin brain weighs approximately sixteen hundred grams, and dolphin brain/body weight ratio is similar to that of humans. Dolphins, like humans, show a high degree of encephalization, the relative enlargement of the cerebral cortex. The dolphins and other toothed whales actually exceed humans and other mammals in terms of the amount of convolution of the cerebral cortex. Convolution is a means of packing more surface area of the brain in a given volume. What it indicates in terms of intelligence is unknown. All we can say is that animals showing more behavioral complexity have more convolutions.

The dolphin brain has evolved differently from that of primates and many other terrestrial mammals. The cetacean cortex did not develop the last stage of cortical evolution found in most living terrestrial mammals; it has retained the basic characteristics of primitive terrestrial and avian forms. However, the dolphin brain developed specialized features not found in the brains of terrestrial mammals. Due to the greater density of water, sound travels about four and a half times faster in the ocean than in air, and the dolphin's brain is adapted for perceiving and rapidly interpreting acoustic information at the requisite speed.

We have observed the dolphins in our research program to spontaneously imitate novel computer-generated whistles. In one case, a dolphin imitated a complex half-second-duration whistle by reproducing the end, then the middle, then the beginning of the sound, all within the following half a second! We are much slower at processing auditory information, and we were unable to perceive these distinctions; only through taping and subsequent inspection of sonograms (visual representations of the whistles) could we discern this behavior.

Dolphins' vision is also a well-developed sense, and has been adapted in that the dolphin has a specialized lens in the eye and a double-slit pupil, which facilitates both underwater and in-air vision, as well as seeing through the water-air interface.

Another unique adaptation for life in the ocean is that dolphins sleep differently than other mammals. Dolphin brains, like other mammalian brains, are composed of two hemispheres. But dolphins, unlike other mammals, sleep in short periods throughout the day and night, with one hemisphere showing typical sleep patterns while the other hemisphere remains alert and able to control breathing and swimming. Each hemisphere rests in turn. In this way the brain can rest and the animal can stay conscious.

In trying to characterize dolphin intelligence in general we rely on behavioral observations from the wild and captivity and look for trends. Dolphins' behavior can be described as exploratory, inquisitive, curious, cautious, altruistic, playful, inventive, and highly involved in maintaining and reaffirming relationships with other members of their social group. Their social behavior is marked by a high degree of interdependence and cooperation.

Observations of these elusive mammals in the field indicate that they maintain elaborate and complex social structures that some investigators have termed "societies." Large groups, often made up of hundreds of dolphins, are usually observed in smaller subgroups of five to eight individuals. These smaller groups seem to be based on sex, age, or kinship factors. Females with young often swim together, and nursery groups have been reported in which the mothers swim in a protective U formation while their youngsters swim inside. Young dolphins leave their mothers after three to six years, with some remaining longer. When females mature and give birth they return to their maternal group, and up to three generations of females have been observed in these affiliations. The strong mother-young bond and unusually long lactation period have been suggested as evidence for acquired learning and social acculturation. There is evidence that young dolphins learn their home ranges and behavior patterns from interactions with their mothers. Young male dolphins leave their mothers and join small groups of juvenile males. Subadult males show tight bonds lasting years and are frequently observed swimming with a few females in their group.

The dolphin's strategy for survival seems to be based on interdependence and cooperation between individuals. For example, dolphins engage in shared watching and guarding of the young, coordinated foraging, coordinated hunting and capture of prey, alarm signals, care-giving behavior as evidenced by physical support or standing by, and defense of other adults. While foraging for food, it has been observed that certain individuals actively participate, while others stay on the periphery of the group. The latter are usually mothers attending their young or another female "aunting" or baby-sitting while the mother is feeding.

Care-giving and altruistic behavior has often been observed between dolphins as well as other toothed cetaceans. Individuals have been seen "standing by" or supporting injured animals from below with their bodies and not eating until the animal has either recovered or died. Throughout history there have been numerous reports of dolphins saving or supporting drowning sailors or swimmers as well.

How do we interpret this behavior? Do these mammals know what they are doing, or is this supportive behavior merely an instinctive or adaptive response to the stimulus of an injured creature? Perhaps the answer lies in the range of plasticity of their response to new situations. For example, Dr. Kenneth Norris, an eminent marine biologist, reported observing the following scene. An adult pilot whale (Globicephala macrorhynchus) was fatally shot, and it began drifting toward the boat of its captors. Two other pilot whales approached it and, rising on either side of the dead animal, pushed it down and away, vanishing below with the dead whale. There have been numerous reported cases in which individual cetaceans have protected individuals of other species of cetacea as well. They have been seen biting harpoon lines trailing from injured animals. This appropriateness and plasticity in behavior strongly suggests that the individuals understand something of the contingencies of the situation. Whether these observed sequences should be considered altruistic or simply care-giving behavior is unclear, but it is a pattern we recognize and hold to be a hallmark of intelligence in our own species.

Another striking example of dolphin intelligence and the plasticity of dolphin behavior comes from observations of dolphin play. Some have argued that intelligence evolves principally due to the need for survival, but certainly new information can be acquired and integrated during leisure time when animals are relatively free from the pressures of daily survival (such as foraging and protection from predators). Captive dolphins living in oceanaria are free of these concerns, and their behavior reflects this, as evidenced by a high degree of inventive play with other animals and objects. In our laboratory we frequently observe dolphins in long periods of toy play and object manipulation using various parts of their bodies. Often two or more toys are manipulated at one time. For example, the young males frequently swim upside down holding a ring in their mouths, or tossing a heavy ball up in the air and catching it. We have also observed the dolphins creating their own toys and then interacting with them. On numerous occasions we have observed dolphins assuming a horizontal position near the pool floor and then jerking their heads up sharply but slightly, releasing what looks like a solid silver ring of air, similar to a smoke ring. The dolphins immediately orient to the ring, and follow it and play

**Figure 1.** Greek coins, circa 480 B.C.E., are an example of humanity's long and close association with the dolphins.

**Figure 2.** The "halo" is a ring of air that this dolphin has snorted out of its blowhole. The dolphin plays with its creation as the ring rises to the surface. *(Photo: Copyright Diana Reiss)*

with it as it rises to the water surface. Sometimes they bite it, or smack it with their tails, or roll their snouts around in it as they would with a solid plastic toy ring. We marveled as we watched a dolphin blow one ring and then another at a faster velocity, which merged with the first ring and formed a hoop-sized ring, which the dolphin swam through. Another day a group of dolphins were observing one dolphin producing air rings when one of the observing animals swam to the pool floor, picked up a piece of fish, and put it into a rising ring. As the ring rose, the piece of fish spun wildly in the turbulence of the torus and the dolphins closely watched and followed it to the surface. Again, I felt I was observing a familiar pattern of exploratory behavior.

Dolphin intelligence has also been investigated by discrimination testing and problem solving in the laboratory. Studies at the University of Hawaii have demonstrated that bottlenose dolphins can learn to comprehend semantic sequences generated by humans in the form of gestures or acoustic signals. The dolphins have shown the capacity for comprehending signals that refer to agents, objects, or actions. (As mentioned previously, other species have also been taught referential codes of communication.) Another study was conducted in the same laboratory in which a dolphin was taught by using food reinforcement to imitate a variety of computer-generated sounds.

In our research program we have been studying the communication and cognitive behavior of a social group of four bottlenose dolphins, two females and their male offspring. The males were born at our laboratory. We took a new approach to understanding their abilities and the way they process information. Rather than using food to train the dolphins to learn certain tasks, we instead wanted to give them some control in obtaining objects and activities and in controlling *our* behavior. We fed the dolphins three times daily and only after they were fed did we work with them. They were provided with an underwater keyboard that displayed various visual forms on the keys. The keyboard was connected to a computer that generated specific whistles underwater when the dolphins pressed particular keys. The use of specific keys also resulted in the dolphins receiving specific things such as balls, rings, floats, or tactile interaction. While all the results cannot be presented here, the most significant finding was that the dolphins spontaneously began to mimic the computer-generated whistles and integrated these new sounds into their own repertoire. Behavioral observations and acoustic recordings indicated that they were using these new sounds in behaviorally appropriate contexts. For example, the "ball" whistle was often recorded as the dolphins were interacting with toy balls during and outside of keyboard sessions. This suggests that the dolphin is capable of making associations between sounds and environmental stimuli and using them in appropriate contexts. There is a significant difference between *training* an animal to do this, which demonstrates its capabilities to do tasks we have designed, and the *spontaneous* development of this behavior by an animal. I suggest that the latter reveals much more about an animal's abilities and information processing. This pattern of association and signal use is a familiar one to us and suggests that there might be similar strategies used for processing, storing, and using information in widely divergent life forms. One thing is clear. We have much more to learn about these remarkable mammals and the extent of their abilities.

BREAKING THE LONG LONELINESS

In observing the behavior of other species we see patterns of behavior that are often familiar, recognizable. These patterns only exist insofar as we are willing to perceive them, however. Let me explain. In 1960 Loren Eiseley wrote an essay called "The Long Loneliness," in which he suggested that intelligence cannot just be conceived of in human terms. As a scientist I realize that we must find the delicate balance between being anthropomorphic, assigning human traits to other animals, and being anthropocentric, assuming that we alone are unique in our abilities and only our kind of intelligence is "real" intelligence. But how do we find this balance in observing complex behavior in other species? We try to apply Lloyd Morgan's law of parsimony: In no case may we interpret an action as the outcome of the exercise of a higher mental faculty if it can be interpreted as the exercise of one that stands lower on the psychological scale. This law of parsimony has been the foundation for scientific investigation for centuries. It is clear that parsimony is generally a good idea, but it can also be a problem if applied in the wrong way. For example, much of our behavior if viewed in this manner could be interpreted as mindless responses to environmental stimuli. There is no clear solution. Rather, the problem challenges us to find new ways of observing and interpreting behavior. In the end it depends on what we are looking for, how we are looking, and what we expect to find. I share the sentiments Eiseley expressed in his poem "Magic":

Ultimately, it is up to the individual to either break or maintain the long loneliness. It is important to realize that in some cases the borders between species are not real but are, rather, assumptions based on a lack of evidence or data. Historically it has been a tacit assumption that we are the only symbol- and tool-using species and are at the pinnacle of evolution. The view that physical evolution is pyramidal has been replaced by a view of evolution as a spreading structure with diverse life forms from different phyla. It is clear that there are both convergent and divergent processes and a variety of strategies operating throughout the biological world that enable different species to survive and flourish in their own environments. Perhaps in the near future we will view the evolution of intelligence in a similar way.

# ALTERNATIVE LIFE DESIGNS

## BY HAL CLEMENT

Scientific opinions, supposedly based on evidence rather than dogma, are tentative by nature. There is no way yet to say with confidence where in the Universe, except on the one planet where we have already found it, there may be life. We can only draw inferences on the matter, as early scientists inferred the shape of the Earth and later ones the existence and nature of the electron.

We may hope that our ideas about life distribution will some day, like our Earth-shape theory, get solid support from further evidence; but so far this is only a hope, dimmed by the fact that any reasoning about extraterrestrial life demands data from the haziest fringes of our knowledge about chemistry and astronomy. I will try in what follows to show where the haziness lies.

The speculation falls into two major parts. We must try to decide what life itself requires for existence, which is fortunately not quite the same as deciding what life is; and we must then settle what environments the Universe offers that may reasonably be expected to provide such needs, and where these may be found. Essentially, we will be discussing chemistry and astronomy.

A warning: Scientists are expected to be objective—that is, to keep their wishes out of their opinions. I am not sure how closely I approach this ideal. I like to suppose that life is extremely common in the Universe, just as at least one television evangelist prefers to believe, for, he says, biblical reasons, that it can exist nowhere but on Earth. You will have to decide for yourself how far my wish has influenced my thinking in sections containing the words, or even the thought, "It seems likely to me that . . ."

Read this essay critically, therefore, and if you are moved afterward to go out and study biology, chemistry, physics, astronomy, or several of them, my feelings will certainly not be hurt.

First, then, life's needs.

All things we regard as living assimilate material from their surroundings and incorporate selected parts of it into their own architecture; they respond in various ways to changes in their environments; they reproduce their kind with more or less accuracy and repair injury to their structures. Some nonliving things, such as fires and crystals growing in solution, show certain of these abilities, but I can think of nothing not alive that has them all.

I do not see how all this can be done without a very complex chemical and mechanical structure. I am leaving the supernatural out of consideration, as science must; this is not because we can disprove it—we cannot—but because we can't work with it. Anyone with a normal imagination can produce an endless number of supernatural explanations for anything, and there is no objective way of choosing among them. Hence, for this essay's reasoning, life involves complex chemistry.

It also involves *solution* chemistry, material in the liquid state. The gas state does not permit definite mechanical structure; if its molecules were close enough to each other for the forces between them to produce a fixed pattern, the substance wouldn't be a gas. A solid, at the other extreme, is too organized; it does not allow atoms and molecules to move around freely enough for the things that have to happen in a living creature. An organism may have solid structural parts, like our bones and teeth, and *handle* gas as we do when we breathe, but its essential chemistry goes on in solution.

Therefore, we need liquid. An easily available liquid.

I don't say it has to be water, though we know water will work. Neither do I suggest that any liquid whatever, such as mercury or lava, would be appropriate. Mercury is quite a rare element on Earth; there is reason to believe that it is comparably rare elsewhere in the Universe, and a planet with enough of this substance to form oceans and lakes, and be available in quantity for life forms, seems pretty unlikely. Melted rock has to be at what we would consider very high temperatures, which implies that much of the complex molecular structure that seems necessary for life could not hold together.

Four materials seem most likely to me, a priori, to be reasonable life solvents. Five others might be added to the list without too much strain on a science-fiction-oriented imagination.

The first four are all hydrogen compounds, which makes them common on theoretical grounds, and three of these are known to be plentiful on various planets we have studied. The fourth is less defensible, perhaps. Hydrogen is observed to be the most common element in the Universe; it appears that about 999 out of every 1,000 atoms that exist are of this simple substance (and plain hydrogen is one of the five others mentioned above).

The elements that combine with hydrogen in my first four suggested liquids are also common; they are the four "hydrogen bonding" atoms of the second period of the chemical table—carbon, nitrogen, oxygen, and fluorine. The liquids are, then, methane, ammonia, water, and hydrogen fluoride.

Liquids? Methane, ammonia, and hydrogen fluoride are all gases under what we consider ordinary conditions. I am claiming, however, that life need not be confined even to such close-to-ordinary conditions as our own ocean-bottom thermal jets, where water, too, would be gas if it weren't for the pressure. All four of these substances are solid if the temperature is low enough, liquid if the temperature is not too high and the pressure is high enough, and gaseous regardless of pressure if the temperature is high enough. Figure 1 gives the approximate ranges. Its vertical scale, temperature, is in Kelvins, used in science because it has no negative—zero Kelvins means there is no heat energy; you can't get less. The bottom section of each column is the temperature range at which the substance is solid; water melts to liquid at 273 Kelvins, for example.

Hydrogen fluoride is the least likely of the four to form oceans and be the common liquid even on a planet with the right temperature and pressure conditions. The trouble is that such worlds are likely to have silicate rock crusts, which are rich in oxygen (there is more oxygen in the top few feet of the solid Earth than in all the air overhead); and hydrogen fluoride will react with such material to form water and silicon tetrafluoride, a gas under what we consider ordinary conditions. The latter, whether gaseous or liquid, can be expected to undergo further reactions that will eventually tie most of the fluorine up in insoluble minerals, just where we find it on Earth.

Water we know about. It forms the basis of Earth's oceans and Earthly life fluids; we know it works as a biological solvent. It has been suggested that it is the only liquid which can, on the basis of some of its really unique properties.

One of the most striking of these is that it expands when it freezes, so that ice floats on liquid water—a rare trait indeed. This prevents oceans and large lakes from freezing all the way to the bottom in winter, thus protecting the life in them from freezing too. I grant the weight of this argument; but on the other hand, the expansion makes things much harder for a water-based life form that actually does get frozen. Its cells tend to be punctured or burst by the growing ice crystals. Try letting an apple or a potato freeze in your ice cube compartment, if nothing of the sort has ever happened around you by accident, and see what the item is like after it thaws out. The expansion argument seems to me to work both ways; other solvents might actually be *better* than water.

The best substitute is probably ammonia, which is a remarkably waterlike compound in spite of its odor. I am not referring to the cleaning liquid sold in grocery stores, which is ordinarily about 2 percent ammonia in water, but the real $NH_3$, which, as Figure 1 indicates, is a gas under normal pressure until we cool it down quite a bit (or a gas at normal temperature unless we squeeze it pretty hard). At low enough temperature and/or high enough pressure, ammonia looks like water, pours and flows like water, and dissolves pretty much the same things in the same way, though not always to the same extent, as water. While details differ, pretty much the same sort of chemistry can go on in ammonia solution as in water solution, as a great deal of laboratory work has shown.

Water and ammonia have molecules whose electrical charges are not smoothly distributed, so they have surface regions distinguishable by positive or negative polarity. This has a profound effect on their ability to dissolve other materials, as well as on their melting and boiling points (hydrogen fluoride is also polar; see how these three compare with all the others on the temperature graph, Figure 1).

Methane, the third of the "most probable" liquids, is not polar. This does not mean that it won't dissolve anything, only that it prefers other nonpolar substances to things like salt. Salts are vital to life as we know it—but so are the highly nonpolar substances we call fats and oils, which our very polar water does not dissolve very well. Living things get around these problems. Nothing is *completely* insoluble in any given liquid, and there are molecules called *detergents* that are chainlike in shape and have one end that is polar and soluble in polar liquids, while the other end is nonpolar and soluble in nonpolar liquids.

Methane's lack of polarity is responsible for its being liquid at much lower temperatures than ammonia, water, or hydrogen fluoride. The five other substances mentioned as possible solvents are, as Figure 1 shows, even more extreme in this respect; and like methane, their dissolving powers would favor nonpolar chemicals.

There are chemicals other than the solvent liquids that go into life architecture, of course. They are, here on Earth, mostly compounds of elements already mentioned: carbon, whose atom can fasten to as many as four others at once and hence be involved in extremely complex structures; the hydrogen-bonders nitrogen and oxygen; and hydrogen itself. One not mentioned before but playing a critical part is phosphorus; nearly as ubiquitous, but not as clearly necessary, is sulfur.

Very tiny amounts of other elements such as cobalt, zinc, molybdenum, and, in at least one type of creature, vanadium, also play active parts in life chemistry. I consider it most unlikely, however, that these were used by the original, simplest organisms that started the biology game on our planet; it seems much more probable to me that life learned to use them later, as children learn to use casually found objects as tools, with natural selection preserving the more effective chemical tricks. I could be wrong, since these elements were present from Earth's beginning; indeed, they are present in the interstellar gas and dust from which the Earth and solar system presumably condensed. If they were *essential* from the beginning, however, life is a much less probable phenomenon than I like to believe, since their concentration is quite low.

The *essential* original life compounds were presumably ones that were simple enough to have formed with fairly high probability during random molecular collisions in solution and then tended to polymerize. That is, they combined with each other to form much more complicated chains of molecules, such as cellulose, which uses only carbon, hydrogen, and oxygen; proteins, requiring the same three plus nitrogen (some of these *now* contain sulfur and iodine as well); and DNA, which also requires phosphorus. Exactly what went on is still being worked out; we don't know.

All these structures depend heavily on the four-bond capacity of carbon to attain their complexity. It has been suggested that other atoms with the same capacity might be used by some exotic life form. The most obvious candidate is silicon, and more recently the silicone group (not the same thing; it has two oxygen atoms as well as a silicon one) has also been suggested. There are three principal, though certainly not conclusive, objections to these ideas. First, the Earth's crust is about a quarter silicon by weight, and much less than one percent carbon; it would seem that if silicon could have been used, it would have been. Second, the four-bond capacity is not the only property of either atom; silicon is noticeably larger than carbon. One familiar result of this size difference is that carbon dioxide is a gas under Earthly conditions, while silicon dioxide (quartz, amethyst, chalcedony, etc.) is not only hard enough to be used for jewelry but is highly insoluble in both polar and nonpolar liquids (remember the importance of the liquid state). Third, silicon does not form *hydrogen bonds*, intermediate-strength chemical links firm enough to keep a human being from collapsing into a puddle of warm jelly but weak enough to permit muscles to flex without preliminary treatment by a blowtorch. Only the second-period atoms carbon, nitrogen, oxygen, and fluorine are small enough to do this.

Figure 2 shows "skeletons" of molecules that meet both the above qualifications and are found as structural members in much of Earthly life. The arms, numbered for convenience on the purine skeleton, indicate points where other atoms or groups of atoms may join them to make more complex units; for example, starting with the purine and adding oxygen atoms to arms 2 and 6, a hydrogen to arm 8, and methyl groups (methane molecules with one hydrogen removed so they also have an "arm") to 1, 3, and 7 gives us caffeine. Adding a hydrogen to the oxygen on 6 and taking the methyl away from 1 turns the caffeine into theobromine, if you prefer chocolate to coffee. Starting back with the skeleton and putting hydrogens at 2, 7, and 8 and an amino group (ammonia with a hydrogen removed) at 6 gives adenine, one of the key bricks in the DNA molecule and also in ATP (adenosine triphosphate, the "battery" involved in the energy reactions of, as far as I am aware, all known life forms on this planet).

We certainly do not know whether these skeletons and the other common ones I haven't listed are the only possible structural items for living beings. It seems most unlikely to me that they are. Earth creatures use only about twenty of the hundreds of possible fairly simple amino acids, and only a few of the possible simple sugars. I am inclined to attribute this to historical accident; groups containing these particular items were the first to acquire the self-replicating ability.

I have not yet written a *Robinson Crusoe* type of space epic, but if I ever do, my marooned hero's chief problem will be finding living tissue, either plant or animal, that he can digest and assimilate. Even on Earth there are lots of kinds we can't handle: Though cotton (cellulose) is a polymer of a simple sugar, and botulism toxin a protein, neither can be processed properly by human digestive enzymes (I didn't use snake venom as an example because Crusoe could digest that with no trouble). I think a Crusoe's chances of finding a native meal would be pretty small on any

# LIQUID RANGES FOR POSSIBLE LIFE SOLVENTS

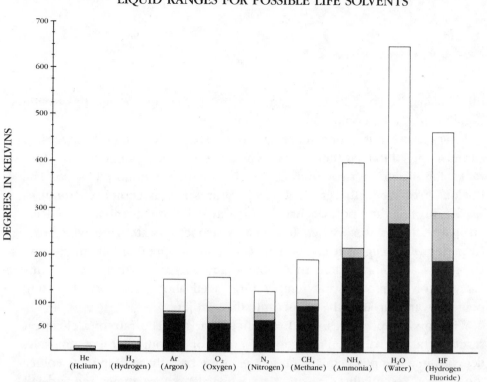

**Figure 1.** The black portion of each bar indicates the temperature range in which each substance is a solid. The shaded portion shows the range at which each substance is a liquid at normal pressure; the clear portion shows the range at which each substance is a liquid at elevated pressure. *(Diagram supplied by Hal Clement. Art by Elizabeth Wen.)*

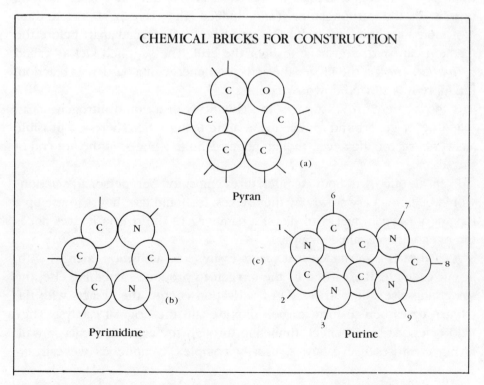

**Figure 2.** The building blocks of life. (a) Pyrans combine with hydrogen and hydroxide to form simple sugars. (b) Pyrimidines and (c) Purines combine with various atoms to form molecules important for life. Numbered bonds are referred to in the text. *(Diagram supplied by Hal Clement. Art by Elizabeth Wen.)*

other planet, even if its temperature were in the right range to provide liquid water and a comfortable oxygen concentration.

Temperature raises one more question. The various liquids suggested earlier imply, if indeed they all will work, a very large possible temperature range for life, from two or three Kelvins up to six hundred or so. The obvious problem with very high temperatures is that complex structures such as proteins and polysaccharides (starch, cellulose) break up or at least change their structure when heated too strongly—watch the white part (protein) of an egg as it cooks. This is not an insuperable difficulty; some structures are more stable than others, and in fact Earth has life forms carrying on comfortably at temperatures well above the normal boiling point of water, around the suboceanic thermal jets.

With low temperatures, another difficulty arises. The rate of a chemical reaction, be it involved in digestion of a meal or in contraction of a muscle, depends in a complicated way on temperature; broadly speaking, the colder the slower. Essentially, a reaction takes place because things tend to fall downhill; that is, to yield to forces. There are forces that tend to fasten carbon atoms to oxygen atoms, forming carbon dioxide; yielding to them is a "fall," and the collision at the end of the fall shows up as heat energy from, say, burning charcoal. However, charcoal does not start burning unassisted; the carbon atoms (and, indeed, the oxygen atoms in their molecules) are stuck to each other and have to be forced apart before the reaction can start; you have to light the grill. The supplied kick is called *activation energy*; the bigger the activation energy of a particular reaction, the more it is slowed down by cooling.

It looks, therefore, as though the chemicals in a liquid nitrogen ocean might never get around to reacting at all in living style. There is a possible answer here, too, however, though not as definite a one as at the hot end of the scale.

Even at our own body temperatures, most of our personal reactions would be going very slowly by themselves. It should take hours to use up a lungful of oxygen, weeks to digest a hamburger. The reason it does not is the phenomenon called catalysis.

A catalyst is a substance that reacts easily (low activation energy) with, in our example, the carbon or the oxygen to produce something else; the something else, still with very low activation energy, then reacts with the oxygen or carbon to form carbon dioxide and the original catalyst. They help reactions by, in effect, tunneling through the activation energy wall. Living creatures use catalysts, usually complex compounds we call *en-*

*zymes*. You use your lungful of oxygen in seconds, digest your hamburger in an hour or two, because you have the appropriate enzymes.

I don't know what enzymes exist that would help out at liquid nitrogen temperatures; I'm not sure there really are or can be any. If there are, I can't be sure that the resulting life speeds would be anywhere near our own, so that creatures with the behavior patterns of Earthly animals would be possible.

To summarize the chemical situation, I consider it likely that ammonia as well as water can act as a reasonable life solvent, and possible though perhaps rather less likely that such nonpolar materials as methane and nitrogen might do so. The structural chemistry of life will probably depend largely on compounds of the second row of the periodic table—carbon, nitrogen, and oxygen—and the ever present hydrogen.

I can't make up my mind how likely it is that phosphorus could be replaced in the DNA ladder or whatever equivalent alien life might use for self-replication. With the size, the charge arrangement, and the availability of the atom all relevant, it is just possible that phosphorus rather than the traditional carbon might prove to be the really essential element for life.

Likely, or fairly likely, temperatures seem to extend from perhaps fifty Kelvins (liquid nitrogen with dissolved impurities; liquid hydrogen and helium strain even my optimism) up to around six hundred or a bit more (near the critical temperature of water, the top of the bar in Figure 1, above which water *must* be gaseous).

The remaining questions are astronomical: Are there many environments providing these chemicals in this temperature range in the universe? Where do we find them? Or, accepting practical restrictions on present science, where would we expect to find them?

The answer to the first question seems to be affirmative. At the moment, astronomers generally feel that planets probably exist near most if not all suns. This has not always been the belief; in my undergraduate days, the accepted theory of planet formation assumed a near collision of stars, a most unlikely event implying that there could be only one or two planetary systems in our entire Milky Way. It is possible that new evidence will force opinion to swing back, of course—science is not a religion—but the present arguments seem much stronger to me than the collision ones ever did (but remember, I *wanted* even in those days to believe in lots of planets). It is therefore reasonable to suppose at the moment that there are billions or hundreds of billions of planets in our Milky Way and hundreds of millions of times as many in the Universe. The questions that remain are how many of these provide environments in which life could exist, and would life come into existence wherever conditions are suitable?

The preceding paragraph's reasoning assumes that planets are associated only with stars. This may not be true. It is easy to imagine cosmic dust accumulating to form small, low-mass objects, just as it is believed to have accumulated to form the larger, more massive ones we call stars, and a Milky Way swarming with sunless planets is quite conceivable, but I am not sure how well actual calculations of the behavior of cosmic clouds shrinking under their own gravity would support the picture. It is hard to see, moreover, how such planets could get warm enough even for liquid nitrogen life unless they somehow accumulated, during formation, rather surprising amounts of uranium, thorium, potassium-40, or other radioactive materials. These elements were not, we believe, present in the earliest clouds of material in space, but were "cooked" in the cores of aging suns and distributed by the "stellar wind" ejected from what astronomers call red giants and by the explosion of supernovas. The earliest stars and planets to form would have contained little but hydrogen and helium, and planets in very old objects such as globular star clusters may be only lifeless gas giants.

The temperature that sunless planets would reach from starlight and other distant radiation sources would be three or four Kelvins at their surfaces. I could, and have, gotten around this in science-fiction stories, but in this more serious discussion we'll leave sunless planets out of consideration.

Stars or suns—the words are synonymous in astronomy—do warm their planets, however; the questions are merely how big and how hot is the star, and how far from it is the planet?

A star is a nuclear reactor whose output depends on its mass; the more material it has, the denser and hotter it is at the center, and the faster its nuclear fusion reaction gulps hydrogen. Our own Sun has, we believe, been shining for about five billion years, and can go on for another three or four billion before the fuel is so thoroughly used up that major changes will occur in its structure (we think it will expand, becoming a red giant, engulfing its inner Earth-type planets before running out of fusible material altogether).

A star twice as massive would have sixteen times the sun's output, and last only about an eighth as long. A sun like Rigel, probably ten times Sol's mass and tens of thousands of times as bright, is unlikely to remain stable for more than ten million years. If it had planets, and they were far enough out and had time enough to cool to the solid state and form oceans, life

## STELLAR TEMPERATURE & BRIGHTNESS

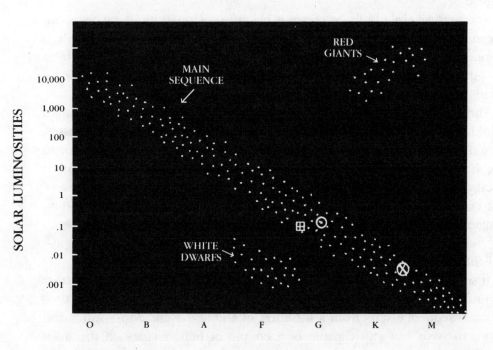

**Figure 3.** A schematic representation of the relationship between temperature and brightness for stars. Our Sun is represented by a dot in a circle. The F star indicated in the text is shown as a boxed plus mark, and the red dwarf is shown as an X in a circle. *(Diagram supplied by Hal Clement. Art by Elizabeth Wen.)*

might come into existence; but it would not have time to evolve very far before the star would roast or swallow the worlds.

In the other direction, a red dwarf like Proxima Centauri—the closest known star outside our planetary system—could simmer along for half a trillion years or so with little change, and if it happened to have a planet close enough for something to be liquid (not necessarily water), there would plenty of time for life to originate and evolve.

Classification of nearby stars has shown that about 90 percent fit on the *main sequence*, which is the line running from upper left to lower right in Figure 3.

The vertical scale in this diagram is brightness (logarithmic; each step up means ten *times* as bright; our Sun equals one unit). The horizontal scale is temperature; the letters have historical significance only, but are still used by astronomers to describe stars. B stars are very hot, M are relatively cool (compared to our Sun, which is a G).

The main sequence, we believe, contains stars in what might be called adulthood, burning their hydrogen at the rate set mainly by their masses. The ones in the upper right, the red giants, are nearly out of hydrogen and are having their senile fling, wasting what they have left at hundreds or thousands of times the Sun's rate. The bottom center ones are white dwarfs, the cooling corpses, still pretty hot by our standards, but out of fuel and no longer kept inflated by nuclear reactions. On the main sequence, mass goes up with brightness as mentioned; off it, the connection is much less reliable.

It seems likely, from the foregoing, that we should hope most reasonably to find the environments mentioned earlier, and hence life, on planets: first, of main-sequence stars from a little left of and above the Sun—hotter and more massive, but still capable of a couple of billion years on the main sequence to give evolution a chance—and on downward to the right; and second, of stars that formed relatively recently from gas and dust that have cycled at least once through earlier suns and contain respectable numbers of heavy atoms. This would leave out globular star clusters, believed to be very ancient; possibly much of the central part of the Milky Way; and the so-called elliptical galaxies. It still leaves in the disc or spiral arm regions of spiral galaxies and many, many billions of stars. My personal opinion is that life, since I believe firmly that it results from the operation of natural laws, *will* come into existence fairly quickly wherever conditions are suitable.

What evidence we have, and there is a good deal, indicates that the number of stars in the Milky Way increases in the low-right direction on the temperature-brightness diagram; that is, there are relatively few of the hot, bright, upper-left members, and enormous numbers of the cooler, dim red dwarfs. There is no lack of stage settings—and, I would be willing to bet, no lack of life.

The hot end for reasonable search planning might be somewhere in the F region, say a star three times as bright as ours, at about the plus mark in Figure 3. It would be 40 percent more massive than the Sun, so should have roughly half or a little less of Sol's eight- or nine-billion-year main-sequence lifetime. This should be ample for evolution. An Earth-like planet would have a temperature of six hundred Kelvins if it was about as far from this star as Earth is from our Sun—what astronomers call an *astronomical unit*. However, for water to be a liquid at six hundred Kelvins, tremendous pressure is needed, and an atmosphere heavy enough to furnish this pressure would also have a strong greenhouse effect (this should not need explaining, these days). The closest habitable planet to our F star would therefore be more like two astronomical units out—perhaps farther; exact greenhouse effects are dependent on detailed atmospheric composition and extremely tedious to calculate—and would have a year of about twenty-eight of our months. The range of possible life, if this essay's arguments are essentially correct, would extend out to Pluto's average distance or thereabouts, where the liquid would be nitrogen and the year a couple of our centuries. This is lots of room.

Toward the faint-star extreme (not *at* it; see Figure 3 again, right-hand "X" mark) would be a red dwarf of about one percent of our Sun's brightness and a quarter of its mass. A planet would have to be within about a tenth of an astronomical unit to receive the same heat flow as we do; such a world would have a year about three weeks long.

A complicating factor in this world's climate would be the likelihood that it would be in tidal lock with its sun, as our moon is with the Earth—that is, one hemisphere would be in perpetual daylight, the other in endless night. With a dense enough atmosphere, heat might still be carried around to the night side fast enough to keep all the gas and liquid from turning solid; the world would now have a really fascinating environment range. Working out the details of such planets is part of the fun of writing careful science fiction. I do not eliminate such a planet as a possible base for life.

By the time we are three astronomical units from our red dwarf, though—about twice as far as Mars is from our own Sun—we are down to the liquid-nitrogen temperature zone. There can easily be planets within such a distance, of course; the system in which we live has four, or five if you count our moon, no farther out than Mars. Any of these *might* have acquired enough material during formation to have a decent atmosphere and a liquid phase. Apparently only two of ours did, and Venus then lost its liquid; but this shows why deciding *how many* life-bearing planets there may actually be is still largely guesswork. I believe the number to be very large indeed; but we need more research on just what occurs on the large scale as planetary systems form, and on the small atomic and molecular one as life comes into being.

I feel quite sure that both sets of phenomena can eventually be explained by the operation of natural laws; but some of you younger folks will have to do a lot of the work. I won't last long enough, probably, to learn the answer, even though discovery is the greatest entertainment I can imagine and like most people I intend to keep enjoying myself as long as I possibly can. Good luck to *you*.

# CHAPTER 2

# SETI THROUGH
# THE AGES

*The search for extraterrestrial intelligence has a long and curious history, dating back at least as far as human experience has been recorded in writing.*

*Thomas R. McDonough shows how humans have sought to make contact with "others," and—perhaps more important—how our efforts have been the reflection of the times and the societies in which they were made. In short, the search for other intelligences has often been a mirror of our own human yearnings and limitations.*

*The ancient Peruvians who drew the mysterious figures across the desert plain of Nazca, for example, may have intended those drawings as meditation aids rather than attempts to contact extraterrestrials. It is our generation that looks at the Nazca figures in the context of SETI. More soberingly, if we cannot tell with any assurance why the Peruvians drew those figures, what can we expect to understand of the workings of totally alien minds, if and when we contact them?*

*Perhaps the best portrayal of a totally alien mind was written by the late Stanley G. Weinbaum in his memorable tale, "A Martian Odyssey." First published in 1934, this intriguing science-fiction story was immediately hailed as a classic, which it remains to this day. The Mars that Weinbaum describes does not really exist; we know that, thanks to the advance of the space sciences. But the Martian character Tweel will stick in your memory even after SETI succeeds and we find real extraterrestrials.*

# ET: PHONE ARISTOTLE!

## BY THOMAS R. McDONOUGH

**T**he sky above the bleak Middle Eastern landscape was shattered by the saucer-shaped vehicle that slowly settled down onto the desert. Zeke, the only human witness, shrieked with terror at the bright flames of its rocket thrusters that dazzled the eye even at midday, while its roar deafened his ears. He threw himself down on the ground in fear.

After the bluish-green saucer had landed, a door slid open, and one by one four space-suited humanoids stepped out onto the ground, placing brass boots carefully onto the sand, staring at the barren land through their faceplates. Emblems of animals were etched onto the sides of their bright red suits.

A loudspeaker on the craft boomed out in the strange tones of the unearthly language. The spacemen unloaded four small helicopters and unfolded their blades. One of them entered a chopper and turned it on. Its roar echoed throughout the valley.

Another alien spotted Zeke and turned on his psychrotronic transmitter-translator. The sounds of the words echoed inside poor Zeke's brain: "On your feet, dude. We have a job for you!"

Science fiction, right? Actually, believe it or not, it's the Bible, freely translated into modern English. It's from the book of Ezekiel, one of the strangest stories in the Old Testament, a story that sounds like a scriptural close encounter of the third kind. It's one of many curious stories that have been passed down from ancient times. Intriguing stories like these make even hard-core skeptics like myself wonder whether perhaps, some time in the distant past, we may have been visited by another civilization.

Since it seems reasonable that there may be other civilizations in the Universe, and since there is no law of physics that prevents interstellar travel, it is logical to ask the question: have we ever been visited in the past? For this chapter, I'll put aside the phenomenon of unidentified flying objects (UFOs) to examine the intriguing question of whether there might be some *historical* record of past visitations. And I'll show the connection between early attitudes about extraterrestrial intelligence and modern efforts to search for it.

Let's look at some of the evidence, from bizarre stories of ancient history up to the interstellar messages that our civilization is sending out in the twentieth century.

## OTHER ANCIENT-ASTRONAUT CANDIDATES

Looking beyond the Bible, we find many intriguing stories and pictures. On the cliffs of the Sahara Desert are found paintings known as the Tassili frescos. As old as 6,000 B.C., these figures look unearthly, but there is good evidence that these are pictures of natives wearing ritual costumes.

Some ancient Mayan carved figures may look a bit like an astronaut about to blast off in an Apollo space capsule, but they clearly portray masked Mayans dressed for rituals. The problem is that masks with exaggerated features of symbolic significance are common among tribal people all over the world, and there is no reason to assume that they necessarily are representations of extraterrestrial beings.

Then there is the case of the Nazca drawings on the low coastal desert of Peru. An extinct civilization drew huge pictures of birds and other creatures on the desert, pictures not evident to the pedestrian, but which are quite visible from the air and, to a limited extent, from the surrounding hills.

They have even been suggested to be landing fields for alien vehicles, though it's not clear why a civilization advanced enough to engage in interstellar travel would need a landing field in the desert. More likely, the Nazcas believed that the gods were up there in the sky. After all, they could see the Sun God rise in the morning, and the Moon God come out at night, and hear the Thunder Gods when they were angry during a storm, flashing their weapons in the sky. It is reasonable that the Nazcas might want to send messages by making big drawings in the sand that the gods above could see easily, such as: "Send more rain, and while you're at it, drop a few stiff lighting bolts down on our enemies over there." Some of the same motifs, probably part of important rituals, are found on Nazca pottery.

The ancients studied the sky because their lives depended on it. During the year, the constellations rose and fell, telling them when to plant and when to harvest. They knew that the five bright planets (Mercury, Venus, Mars, Jupiter, Saturn) were special because they wandered slowly among the thousands of "fixed" stars, although until the telescope arrived, they couldn't know that these were also worlds like ours. Occasionally, a star would appear mysteriously, and this was cause for great worry, because it meant that something was dramatically disrupting the slow, predictable backdrop of the gods. It was taken as an omen that something major was about to happen, or perhaps it was signaling something that had just occurred: the death of a king, war, famine. Today, we know these apparitions were actually supernovas (exploding stars) and comets (huge chunks of dirty ice flying between the planets).

What I look for in claims of ancient extraterrestrial visitations is some evidence from the old accounts that ancient people knew something that they could not have known using the science of their time. For example, one of the most fundamental of all numbers in science is the speed of light. It's something that cannot be measured except by telescopic observations of Jupiter's moons or with some moderately advanced technology. So if I came across some ancient record stating that the speed of light was so many cubits per heartbeat, and if that number were roughly equivalent to 300,000 kilometers per second, I would be extremely impressed. Yet I know of no such revelation.

Still, intriguing tales abound in history. Carl Sagan, in the book he wrote with the Soviet astrophysicist I. S. Shklovskii, *Intelligent Life in the Universe,* disclosed a remarkable tale from the ancient Persian Gulf suggesting contact between humans and extraterrestrials. He quoted an ancient Babylonian account:

"In the first year there made its appearance from a part of the Persian Gulf which bordered upon Babylonia, an animal endowed with reason, who was called Oannes. [According to the account of Apollodorus] the whole body of the animal was like that of a fish; it had under a fish's head another head and also feet below, similar to those of a man, except joined to the fish's tail. His voice too and language was articulate and human.... This Being, in the day-time used to converse with men but took no food at that season; and he gave them an insight into letters and sciences and every kind of art. He taught them to construct houses, to found temples, to compile laws, and explained to them the principles of geometrical knowledge."

But every human society invents myths of its origins, usually a mixture of history and fantasy. Tempting though it is to interpret these as extra-terrestrial visitations, we have to look for hard evidence. If we ever find a laser buried, undisturbed, underneath a Babylonian temple, we'll know either that the ancients were really visited by aliens, or someone has planted a hoax. Up to now, no one has found any artifact that has been studied by independent laboratories that couldn't have been built right here on Earth.

A fascinating modern case sometimes interpreted as extraterrestrial con-tact is that of the African tribe called the Dogon. Living in Mali, the Dogon people have an oral tradition filled with several interesting bits of astro-nomical lore, information that would seem to have been unavailable to them until recently. As reported in Robert K. G. Temple's book *The Sirius Mystery,* the Dogon claim to have been visited by beings from the star Sirius, and they say that that star has a companion that moves around it every fifty years. The amazing fact is that Sirius has a companion invisible to the naked eye that indeed orbits around it every fifty years, and this companion—a white dwarf—was not discovered until the nineteenth cen-tury. Furthermore, the Dogon describe that companion as being the small-est and heaviest object in the sky, and while that is not absolutely true, it is correct that a white dwarf star is vastly denser than an ordi-nary star.

Fascinating and impressive, but science writer Ian Ridpath, writing in the journal *Skeptical Inquirer,* finds all kinds of flaws in the argument. He points out that the primary star Sirius has such a short lifetime that life could not have survived there for very long, and probably could never have arisen there. Furthermore, before it became a white dwarf, the companion was presumably a red giant star and perhaps a nova (a star which blew off its outer layers). Both stages would destroy any nearby planets. It is a hostile environment, nothing like the one around the mild-mannered, sta-ble star we are fortunate enough to have in our own system. And he finds that the Dogon believe that Sirius had *two* companions, not one. Astrono-mers find only one companion.

According to Ridpath, it turns out that the Dogon really say that the companion's orbit is one hundred years, not fifty, and they celebrate the renewal of the world every sixty years, not fifty. And the natives do not actually claim that they were visited from Sirius. They only make vague statements about Sirius that, with a lot of optimism, can be interpreted as implying that they have been influenced by visitors from Sirius.

**Figure 1.** Taken from a rock fresco at the Tassili site in the Sahara, this image has been mistaken for an alien in a space suit by Erich von Daniken and others. It is almost certainly an illustration of a local inhabitant in ritual garb.

**Figure 2.** The curious lines inscribed on the Plain of Nazca are also found as a motif in local pottery. It must have had a significance to the local inhabitants, probably of a religious nature. *(Photo: William R. Alschuler.)*

DUELING PHILOSOPHERS

Whether or not our ancestors were visited by aliens, the possibility that other worlds could exist was one that many of them pondered. The ancients, in those days before the telescope, had only a few paths open to them for their speculation about the Universe: They could extrapolate from the example of the Earth and suppose that there could be other Earths. They could use philosophy and conclude, as Aristotle did, that this must be the only world. Or they could rely on faith, and conclude that because the Bible said nothing about other worlds, that there could not be other worlds.

The concept of worlds other than the Earth existing and even being populated is one that has emerged from diverse philosophies and religions from ancient times to present. Often rejected in the end, it has come to dominate our thinking. In 400 B.C., the Greek philosopher Metrodorus of Chios wrote, "It is unnatural in a large field to have only one shaft of wheat and in the infinite Universe only one living world."

Aristotle's reasoning to the contrary dominated Western civilization for two thousand years. Astronomers and philosophers have had to fight a battle against the rigidity of his reasoning. Just as thinkers had to fight for centuries to overcome the idea that the Earth was the center of the Universe—because to think otherwise diminished the standing of human beings—they also had to fight to support the idea that there could be other worlds in the Universe. This too diminished man's standing. It meant that our planet might not be the noblest in the Universe.

Some scientists paid dearly for their opposing beliefs. Galileo was forced to recant by the Church, his books were banned, and he died under house arrest. One heretical Italian went so far in his thinking as to earn the ultimate penalty. Giordano Bruno, a contemporary of Galileo, was a philosopher who questioned many of the beliefs of the Church in the sixteenth century. He was convinced that there was an infinite number of worlds in the Universe, which contradicted the Aristotelian ideas that the Church had embraced. This was only one of Bruno's many heresies, and in the end the Church paid him the ultimate compliment for his intellectual daring by burning him at the stake in the year 1600.

The invention of the telescope in 1608 opened up the Universe. Now the blind could see. It would allow the planets to be revealed as worlds like ours, and new planets were found: Uranus, Neptune, Pluto. Might there be more? The telescope allowed us to see that the Milky Way was made up of multitudes of stars. It eventually enabled us to see that some of the faint fuzzy objects visible to the naked eye at night were actually made of stars, too—groups we now call galaxies. And it showed that the Milky Way was but one out of many galaxies.

With each major step in astronomy, we have learned that the Universe is larger than we had thought it. And we have learned that there is nothing obviously unique here. The Sun is an ordinary star, just like billions of others in our galaxy. The chemicals out of which humans are made are widely distributed, not rare, in the Universe. And there is evidence that planets are a normal part of the evolution of many stars. All of this suggests today that life may be common in the Universe.

The great seventeenth-century astronomer Kepler believed that there were inhabited worlds in the sky. Kepler, the man who proved that planets move in elliptical orbits about the Sun, even "deduced" the existence of intelligent beings on our moon. He noted a large circular feature that Galileo had observed, which we now know is just a crater. But Kepler asserted that it was created by a civilization of beings who "make their homes in numerous caves hewn out of that circular embankment." In this error, Kepler pioneered the type of reasoning that has led to numerous claims in later times that indistinct objects barely visible through tele-scopes, vaguely resembling human artifacts, on our moon and Mars, proved the existence of intelligent life there—arguments that continue to this day.

In the seventeenth century, the great French philosopher and mathematician René Descartes was able to write, "It seems to me that the mystery of the incarnation and all the other advantages which God bestowed on man do not preclude the possibility that he might have granted infinitely many others, very great, to an infinity of other creatures." But there were many who opposed Descartes's views. One such opponent, Dutch philosopher Gerhard DeVries, wrote, "There are indeed many and weighty authors who would stand in the battle line for the lunar inhabitants. Nevertheless, victory is certainly ours for they are not easily armed by fate or reason."

Alexander Pope's "Essay on Man" summed up the attitude of the eighteenth century's most daring thinkers:

> He, who through vast immensity can pierce
> See worlds on worlds compose one universe,
> Observe how system into system runs,
> What other planets circle other suns,
> What varied Being peoples every star,
> May tell why Heaven has made us as we are.

PHONING ET

All these ideas prepared civilization so well for the idea of life on other worlds that many people in 1835 were taken in when the *New York Sun* ran a series of newspapers articles about the famous British astronomer Sir John Herschel, who was observing the sky from a new telescope in South Africa. The paper claimed that the scientist had seen creatures on the moon, and included drawings of critters who looked just like apes. They had "short and glossy copper-colored hair and had wings composed of thin membranes ... They are doubtless innocent and happy creatures, not withstanding some of their amusements would but ill comport with our terrestrial notions of decorum." The poor astronomer had no idea that reporter Richard Locke was inventing completely phony stories about Herschel's legitimate observations. The Moon Hoax suckered in half of New York City and, amazingly, many people continued to believe in it even after the reporter confessed his crime.

In the 1820s, the German mathematician Karl Friedrich Gauss had suggested that we could announce our existence to the Universe for the benefit of other civilizations that might be out there. He wanted to plant trees in a large area in Siberia in the shape of a right triangle to demonstrate Pythagoras's theorem. This would prove to them that we were at least smart enough to pass high school geometry! As modern SETI researcher Frank Drake likes to say, the result of this proposal was an echo of so many research proposals today: "Not Funded."

The greatest boon to nineteenth-century public interest in SETI was American astronomer Percival Lowell, whose startling "detection" of Martians had long-lasting effects on astronomy and science fiction. It all started in 1877 when he learned that an Italian astronomer, Giovanni Schiaparelli, had reported seeing *canali* on Mars. Schiaparelli just meant *channels*—not necessarily artificial features—but Lowell thought he meant *canals,* which required a civilization to build them.

Sure enough, when Lowell looked through his telescope at Mars, he thought he could see canals. Thus began the century-long hunt for the canals of Mars and the civilization that built them, a search that was to end only when spacecraft flew by the planet and photographed the surface. The problem is that Mars is covered with dark splotches that change from month to month with the Martian seasons. And the difficulty is that when you look through the soupy atmosphere of the Earth at distant Mars, the eye connects these blotches with straight lines. (Photos don't capture this effect very well.) Thus Mars sometimes appears to be crisscrossed by these faint straight lines, and Lowell and other astronomers inspired by him made elaborate maps of these "canals."

To Lowell, these features meant that a civilization must have built the canals to bring water from the Martian ice caps (since confirmed to be real, made of water and carbon dioxide). He supposed that the civilization would use the canals to support crops in warmer regions, and that these crops were the splotches that varied seasonally. It was a magnificent theory, although one that not every astronomer of the time agreed with. But his books about the Martian civilizations were enormously popular, leading to the public fascination with Mars, and inspiring H. G. Wells' novel, *The War of the Worlds*, and countless imitations.

Many other astronomers were cautious about Martians. Sir Robert Ball was such a one. In 1893, he wrote:

"That there may be types of life of some kind or other on Mars is, I should think, very likely ... Speculations have also been made as to the possibility of there being intelligent inhabitants on this planet, and I do not see how anyone can deny the possibility, at all events, of such a notion ...

"We have also heard surmises as to the possibility of the communication of interplanetary signals between the Earth and Mars, but the suggestion is a preposterous one. Seeing that a canal 60 miles wide and 1,000 miles long is an object only to be discerned on exceptional occasions and under most favorable circumstances, what possibility would there be that, even if there were inhabitants on Mars who desired to signal this Earth, they could ever succeed in doing so?"

French astronomer Camille Flammarion had speculated in 1892 that, in the distant future, civilization might find a way of communicating across interplanetary space. But he said it would require some completely new invention. What he didn't realize was that the invention of which he spoke wouldn't take centuries to arrive—it was right there on Earth at that very moment. The fantastic new technology was radio.

Sir Robert Ball not only rejected optical communication, he also pooh-poohed the idea of interplanetary radio communication. He said that one would need a transmitter "sixteen million times as efficient" as radio's "most honored champions" predicted. It would have blown Sir Robert's mind to learn that we now communicate over billions of miles to our spacecraft at the edge of the solar system. Unfortunately, he was born too early to have heard of Arthur C. Clarke's First Law: "When a distinguished but elderly scientist states that something is possible, he is almost certainly right. Whenever he states that something is impossible, he is very probably wrong."

Radio actually turns out to be the best method for SETI (according to current thinking), and radio pioneers Nikola Tesla and Guglielmo Marconi were quick to listen for Martian radio signals. Tesla was the brilliant and eccentric Yugoslavian inventor who fought with Marconi over credit for the invention of radio. Tesla has become something of a cult figure today, complete with a rock band named after him half a century after his death. He was confident that life existed on other planets, especially Mars.

In 1901, Tesla recorded that he had detected strange signals on his most powerful receiver at Colorado Springs. He concluded that the regular pattern of these rhythmic signals must be produced by intelligent life on other worlds, attempting to communicate with Earth, and he proposed that Venus or Mars were the most probable sources. History later repeated itself in 1967, when scientists in England detected rhythmic radio sounds. Those signals in fact were from space, and the scientists at first thought that they too had detected another civilization. But we now know that the 1967 sounds were from pulsars, spinning neutron stars (supernova remnants), not other civilizations.

In 1902, the distinguished British physicist Lord Kelvin visited the United States and announced that he agreed with Tesla that Mars was signaling Earth. The Englishman said that New York was the "most marvelously lighted city in the world," and that it could be the only place on the Earth visible to the Martians. He said further that "Mars is signaling . . . to New York."

By the Roaring Twenties, radio was booming commercially. Marconi also thought he had detected signals from Mars. In 1924, during a close approach of Mars, the U.S. military conducted a search for radio signals from Mars, even assigning a code expert to decode the Martian signals. One civilian observer detected radio signals recorded on a moving strip of film and concluded that they represented a human face. (Not the last time that people would claim to have detected human faces on Mars.) It was actually terrestrial radio interference, probably produced by the growing number of human radio stations, some of whose signals bounced all the way around the planet, reflected by the ionosphere.

The strange radio signals that Tesla and Marconi detected from time to time could have come from a number of sources. Civilization had learned to use powerful electrical machinery, which often produced radio noise as a by-product, especially when sparks were created. Automobile ignitions or the running of an elevator motor could have generated such regular, artificial noises. It is also likely that some of the early investigators were fooled by natural signals called whistlers, produced by distant lightning, that sound artificial.

This concern with Martians came to a peak in 1938, when Orson Welles broadcast what was to become the most famous radio show in history. He did an updated version of H. G. Wells's *War of the Worlds,* and he did it in an unprecedented docudrama style, as if it were an ordinary broadcast interrupted by news bulletins from reporters in the field. To an audience of millions, the radio announced invasion and destruction by the evil Martians as they rampaged across the American countryside, viciously slaying all humans they encountered. Thousands of listeners panicked, and for a few hours SETI was transformed for some of these people to a search over their shoulder for Martians coming down the street.

## MARTIANS MEET THE SPACE AGE

Throughout much of the last century, astronomers have been so embarrassed by the exaggerated claims of the detection of other civilizations that they became gun-shy about the idea of life on other worlds. In the first half of the twentieth century, the idea of extraterrestrial life fell into disrepute in the eyes of scientists, while it flourished in the genre of science fiction. It wasn't until Sputnik was launched in 1957 that ideas in the science-fictional trash heap, such as space travel and alien beings, began to return to respectability.

That started a long uphill battle for serious SETI undertaken by scientists such as Giuseppe Cocconi, Philip Morrison, Frank Drake, and Carl Sagan. They marshaled arguments from what we now knew about the origin of life on Earth and what astronomy revealed about conditions elsewhere in the Universe. Gradually, they convinced much of the scientific community that the idea of life on other worlds was respectable.

Ironically, there has now been a counterrevolution in the form of a handful of scientists who think that we are, after all, alone in the Universe. Scientists, of whom the most outspoken is Frank Tipler, have produced an array of arguments that conditions on Earth are truly unique and that life simply could not arise elsewhere in the Universe. So, strangely, a few scientists have returned to the days of Aristotle, when philosophers sitting at their desks could conclude that an entire vast Universe, about which we still know so little, is empty of extraterrestrial life.

To most scientists this is an absurd conclusion. A Universe empty of anyone but us is conceivable, but it's too early to jump to such a conclusion. After all, we have barely begun to search the skies with a handful of telescopes, at a few frequencies, in a limited number of programs, looking for evidence of other possible civilizations. And so many of the modern discoveries of the twentieth century seem to point in the direction of life elsewhere: the ease of producing organic molecules under conditions like those on primitive Earth; dust rings around nearby stars similar to those that led to the formation of the planets; the probable detection of planets around other stars; and the discovery of vast clouds of chemicals in interstellar space—the same chemicals out of which we are made, such as water, carbon monoxide, alcohol, and many others.

While Lowell left a bad taste about ETs in the mouths of scientists, we still owe him a debt of gratitude. Not only did he stimulate public interest in astronomy, but he even indirectly inspired one of the most remarkable space projects of the twentieth century, the Viking missions to Mars. The question of whether there was *life* on Mars lingered long after astronomers relegated to embarrassed silence the speculations of a civilization there. This fascination with the red planet led to a continuing series of space probes, something that probably would have been even more astounding to Lowell than actual Martians.

The first successful one was Mariner 4, which flew by Mars in 1965. It revealed a bleak, cratered landscape showing no signs of canals, one that seemed to have no potential for life of any kind, much less an advanced civilization. The surface looked like our moon. Later spacecraft changed our attitudes once more. They showed that there were huge canyons apparently cut by water on Mars. So it turns out that Schiaparelli was right after all. There are channels on Mars. But not canals.

The fascination with Mars continued, supported by the fact that it was the nicest place in the solar system for life as we know it (apart from Earth). Thus NASA sent two Viking spacecraft to land on Mars. The experiments on board tried to detect chemical reactions due to life, but gave ambiguous results. The concensus is that they did not detect life at the two barren points examined. However, both the U.S. and Soviet space programs have ambitious plans to send more advanced spacecraft there, culminating most likely in a human visit.

And not only have interplanetary spacecraft left Earth, but the first *interstellar* spacecraft have been launched. In 1972, Pioneers 10 and 11 were launched toward Jupiter and Saturn on trajectories that would eventually shoot them out of the solar system. In 1983, Pioneer 10 became the first human hardware ever to cross the orbits of Pluto and Neptune and leave the Sun forever.

Meanwhile, two Voyager spacecraft were launched in the footsteps of the Pioneers. These were bigger, more advanced probes also designed to fly by Jupiter and Saturn, with Voyager 2 destined to add Uranus and Neptune to its list of worlds explored. Like the Pioneers, they will eventually escape the Sun's gravity completely, and wander endlessly through the Galaxy.

Taking advantage of this, NASA put messages on all four spacecraft, designed to be understandable by an alien civilization. The Pioneers each have a plaque of a man and woman, together with a little astronomical information. The Voyagers bear a more sophisticated message, a kind of videodisk with pictures, human speech, and music. It would take them thousands of centuries to travel to the nearest star if they were aimed at it (which they're not). But one day, perhaps billions of years from now, some creature may find a drifting, pitted spacecraft, and know that two-legged creatures built it.

## THE FUTURE

So where do we stand now? We are fortunate to live in the first period in the history of this planet when we may be able to answer the questions that theologians, philosophers, and scientists have fruitlessly argued for thousands of years. With our observatories and spacecraft we at last have the capability of detecting some kinds of life if it is out there.

There is presently no convincing evidence that Earth has been visited at any time in the past—at least, no evidence compelling to the majority of scientists. There's always the possibility that some paleontologist may find an interstellar lunch box mingled with the bones of a long-dead dinosaur. Or perhaps deep in the archives of the British Museum there may be some ancient cuneiform tablet bearing the equation $E = mc^2$, just waiting to be deciphered by a patient scholar.

While there are intriguing mysteries—and one can never completely rule out the possibility that aliens may have interfered with human history—most cases put forth in sensationalistic books like *Chariots of the Gods?* turn out to have been produced by our ancestors (or by the author), not alien visitors. It is the *lack* of imagination in the authors of such books that creates most of the mystery: they are not as smart as our ancestors.

But the next few years may see the answer to the ancient debate of the Greeks. We may finally know whether we live in a Universe as one shaft of wheat in an otherwise empty desert, or whether, as many scientists suspect, we are in a vast field covered with life as far as the telescope can see.

# "A MARTIAN ODYSSEY"

## BY STANLEY G. WEINBAUM

Jarvis stretched himself as luxuriously as he could in the cramped general quarters of the *Ares.*

"Air you can breathe!" he exulted. "It feels as thick as soup after the thin stuff out there!" He nodded at the Martian landscape stretching flat and desolate in the light of the nearer moon, beyond the glass of the port.

The other three stared at him sympathetically—Putz, the engineer, Leroy, the biologist, and Harrison, the astronomer and captain of the expedition. Dick Jarvis was chemist of the famous crew, the *Ares* expedition, first human beings to set foot on the mysterious neighbor of the earth, the planet Mars. This, of course, was in the old days, less than twenty years after the mad American Doheny perfected the atomic blast at the cost of his life, and only a decade after the equally mad Cardoza rode on it to the moon. They were true pioneers, these four of the *Ares.* Except for a half-dozen moon expeditions and the ill-fated de Lancey flight aimed at the seductive orb of Venus, they were the first men to feel other gravity than earth's, and certainly the first successful crew to leave the earth-moon system. And they deserved that success when one considers the difficulties and discomforts—the months spent in acclimatization chambers back on earth, learning to breath the air as tenuous as that of Mars, the challenging of the void in the tiny rocket driven by the cranky reaction motors of the twenty-first century, and mostly the facing of an absolutely unknown world.

Jarvis stretched and fingered the raw and peeling tip of his frost-bitten nose. He sighed again contentedly.

"Well," exploded Harrison abruptly, "are we going to hear what happened? You set out all shipshape in an auxiliary rocket, we don't get a peep for ten days, and finally Putz here picks you out of a lunatic ant-heap with a freak ostrich as your pal! Spill it, man!"

"Speel?" queried Leroy perplexedly. "Speel what?"

"He means *'spiel'*," explained Putz soberly. "It iss to tell."

Jarvis met Harrison's amused glance without the shadow of a smile. "That's right, Karl," he said in grave agreement with Putz. *"Ich spiel es!"* He grunted comfortably and began.

"According to orders," he said, "I watched Karl here take off toward the North, and then I got into my flying sweat-box and headed South. You'll remember, Cap—we had orders not to land, but just scout about for points of interest. I set the two cameras clicking and buzzed along, riding pretty high—about two thousand feet—for a couple of reasons. First, it gave the cameras a greater field, and second, the under-jets travel so far in this half-vacuum they call air here that they stir up dust if you move low."

"We know all that from Putz," grunted Harrison. "I wish you'd saved the films, though. They'd have paid the cost of this junket; remember how the public mobbed the first moon pictures?"

"The films are safe," retorted Jarvis. "Well," he resumed, "as I said, I buzzed along at a pretty good clip; just as we figured, the wings haven't much lift in this air at less than a hundred miles per hour, and even then I had to use the under-jets.

"So, with the speed and the altitude and the blurring caused by the under-jets, the seeing wasn't any too good. I could see enough, though, to distinguish that what I sailed over was just more of this grey plain that we'd been examining the whole week since our landing—same blobby growths and the same eternal carpet of crawling little plant-animals, or biopods, as Leroy calls them. So I sailed along, calling back my position every hour as instructed, and not knowing whether you heard me."

"I did!" snapped Harrison.

"A hundred and fifty miles south," continued Jarvis imperturbably, "the surface changed to a sort of low plateau, nothing but desert and orange-tinted sand. I figured that we were right in our guess, then, and this grey plain we dropped on was really the Mare Cimmerium which would make my orange desert the region called Xanthus. If I were right, I ought to hit another grey plain, the Mare Chronium, in another couple of hundred miles, and then another orange desert, Thyle I or II. And so I did."

"Putz verified our position a week and a half ago!" grumbled the captain. "Let's get to the point."

"Coming!" remarked Jarvis. "Twenty miles into Thyle—believe it or not—I crossed a canal!"

"Putz photographed a hundred! Let's hear something new!"

"And did he also see a city?"

"Twenty of 'em, if you call those heaps of mud cities!"

"Well," observed Jarvis, "from here on I'll be telling a few things Putz didn't see!" He rubbed his tingling nose, and continued, "I knew that I had sixteen hours of daylight at this season, so eight hours—eight hundred miles—from here, I decided to turn back. I was still over Thyle, whether I or II I'm not sure, not more than twenty-five miles into it. And right there, Putz's pet motor quit!"

"Quit? How?" Putz was solicitous.

"The atomic blast got weak. I started losing altitude right away, and suddenly there I was with a thump right in the middle of Thyle! Smashed my nose on the window, too!" He rubbed the injured member ruefully.

"Did you maybe try vashing der combustion chamber mit acid sulphuric?" inquired Putz. "Sometimes der lead giffs a secondary radiation—"

"Naw!" said Jarvis disgustedly. "I wouldn't try that, of course—not more than ten times! Besides, the bump flattened the landing gear and busted off the under-jets. Suppose I got the thing working—what then? Ten miles with the blast coming right out of the bottom and I'd have melted the floor from under me!" He rubbed his nose again. "Lucky for me a pound only weighs seven ounces here, or I'd have been mashed flat!"

"I could have fixed!" ejaculated the engineer. "I bet it vas not serious."

"Probably not," agreed Jarvis sarcastically. "Only it wouldn't fly. Nothing serious, but I had my choice of waiting to be picked up or trying to walk back—eight hundred miles, and perhaps twenty days before we had to leave! Forty miles a day! Well," he concluded, "I chose to walk. Just as much chance of being picked up, and it kept me busy."

"We'd have found you," said Harrison.

"No doubt. Anyway, I rigged up a harness from some seat straps, and put the water tank on my back, took a cartridge belt and revolver, and some iron rations, and started out."

"Water tank!" exclaimed the little biologist, Leroy. "She weigh one-quarter ton!"

"Wasn't full. Weighed about two hundred and fifty pounds earth-weight, which is eighty-five here. Then, besides, my own personal two hundred and ten pounds is only seventy on Mars, so, tank and all, I grossed a hundred and fifty-five, or fifty-five pounds less than my every-day earth-weight. I figured on that when I undertook the forty-mile daily stroll. Oh—of course I took a thermo-skin sleeping bag for these wintry Martian nights.

"Off I went, bouncing along pretty quickly. Eight hours of daylight meant twenty miles or more. It got tiresome, of course—plugging along over a soft sand desert with nothing to see, not even Leroy's crawling biopods. But an hour or so brought me to the canal—just a dry ditch about four hundred feet wide, and straight as a railroad on its own company map.

"There'd been water in it sometime, though. The ditch was covered with what looked like a nice green lawn. Only, as I approached, the lawn moved out of my way!"

"Eh?" said Leroy.

"Yeah, it was a relative of your biopods. I caught one—a little grasslike blade about as long as my finger, with two thin, stemmy legs."

"He is where?" Leroy was eager.

"He is let go! I had to move, so I plowed along with the walking grass opening in front and closing behind. And then I was out on the orange desert of Thyle again.

"I plugged steadily along, cussing the sand that made going so tiresome, and, incidentally, cussing that cranky motor of yours, Karl. It was just before twilight that I reached the edge of Thyle, and looked down over the grey Mare Chronium. And I knew there was seventy-five miles of *that* to be walked over, and then a couple of hundred miles of that Xanthus desert, and about as much more Mare Cimmerium. Was I pleased? I started cussing you fellows for not picking me up!"

"We were trying, you sap!" said Harrison.

"That didn't help. Well, I figured I might as well use what was left of daylight in getting down the cliff that bounded Thyle. I found an easy place, and down I went. Mare Chronium was just the same sort of place as this—crazy leafless plants and a bunch of crawlers; I gave it a glance and hauled out my sleeping bag. Up to that time, you know, I hadn't seen anything worth worrying about on this half-dead world—nothing dangerous, that is."

"Did you?" queried Harrison.

"*Did I!* You'll hear about it when I come to it. Well, I was just about to turn in when suddenly I heard the wildest sort of shenanigans!"

"Vot iss shenanigans?" inquired Putz.

"He says, 'Je ne sais quoi,'" explained Leroy. "It is to say, 'I don't know what.'"

"That's right," agreed Jarvis. "I didn't know what, so I sneaked over to find out. There was a racket like a flock of crows eating a bunch of canaries—whistles, cackles, caws, trills, and what have you. I rounded a clump of stumps, and there was Tweel!"

"Tweel?" said Harrison, and "Tveel?" said Leroy and Putz.

"That freak ostrich," explained the narrator. "At least, Tweel is as near as I can pronounce it without sputtering. He called it something like 'Trrrweerrlll.' "

"What was he doing?" asked the captain.

"He was being eaten! And squealing, of course, as any one would."

"Eaten! By what?"

"I found out later. All I could see then was a bunch of black ropy arms tangled around what looked like, as Putz described it to you, an ostrich. I wasn't going to interfere, naturally; if both creatures were dangerous, I'd have one less to worry about.

"But the bird-like thing was putting up a good battle, dealing vicious blows with an eighteen-inch beak, between screeches. And besides, I caught a glimpse or two of what was on the end of those arms!" Jarvis shuddered. "But the clincher was when I noticed a little black bag or case hung about the neck of the bird-thing! It was intelligent! That or tame, I assumed. Anyway, it clinched my decision. I pulled out my automatic and fired into what I could see of its antagonist.

"There was a flurry of tentacles and a spurt of black corruption, and then the thing, with a disgusting sucking noise, pulled itself and its arms into a hole in the ground. The other let out a series of clacks, staggered around on legs about as thick as golf sticks, and turned suddenly to face me. I held my weapon ready, and the two of us stared at each other.

"The Martian wasn't a bird, really. It wasn't even bird-like, except just at first glance. It had a beak all right, and a few feathery appendages, but the beak wasn't really a beak. It was somewhat flexible; I could see the tip bend slowly from side to side; it was almost like a cross between a beak and a trunk. It had four-toed feet, and four-fingered things—hands, you'd have to call them, and a little roundish body, and a long neck ending in a tiny head—and that beak. It stood an inch or so taller than I, and—well, Putz saw it!"

The engineer nodded. *"Ja!* I saw!"

Jarvis continued. "So—we stared at each other. Finally the creature went into a series of clackings and twitterings and held out its hands toward me, empty. I took that as a gesture of friendship."

"Perhaps," suggested Harrison, "it looked at that nose of yours and thought you were its brother!"

"Huh! You can be funny without talking! Anyway, I put up my gun and said 'Aw, don't mention it,' or something of the sort, and the thing came over and we were pals.

"By that time, the sun was pretty low and I knew that I'd better build a fire or get into my thermo-skin. I decided on the fire. I picked a spot at the base of the Thyle cliff, where the rock could reflect a little heat on my back. I started breaking off chunks of this desiccated Martian vegetation, and my companion caught the idea and brought in an armful. I reached for a match, but the Martian fished into his pouch and brought out something that looked like a glowing coal; one touch of it, and the fire was blazing—and you all know what a job we have starting a fire in this atmosphere!

"And that bag of his!" continued the narrator. "That was a manufactured article, my friends; press an end and she popped open—press the middle and she sealed so perfectly you couldn't see the line. Better than zippers.

"Well, we stared at the fire a while and I decided to attempt some sort of communication with the Martian. I pointed at myself and said 'Dick'; he caught the drift immediately, stretched a bony claw at me and repeated 'Tick.' Then I pointed at him, and he gave that whistle I called Tweel; I can't imitate his accent. Things were going smoothly; to emphasize the names, I repeated 'Dick,' and then, pointing at him, 'Tweel.'

"There we stuck! He gave some clacks that sounded negative, and said something like 'P-p-p-root.' And that was just the beginning; I was always, 'Tick,' but as for him—part of the time he was 'Tweel,' and part of the time he was 'P-p-p-proot,' and part of the time he was sixteen other noises!

"We just couldn't connect. I tried 'rock,' and I tried 'star,' and 'tree,' and 'fire,' and Lord knows what else, and try as I would, I couldn't get a single word! Nothing was the same for two successive minutes, and if that's a language, I'm an alchemist! Finally I gave it up and called him Tweel, and that seemed to do.

"But Tweel hung on to some of my words. He remembered a couple of them, which I suppose is a great achievement if you're used to a language you have to make up as you go along. But I couldn't get the hang of his talk; either I missed some subtle point or we just didn't *think* alike—and I rather believe the latter view.

"I've other reasons for believing that. After a while I gave up the language business, and tried mathematics. I scratched two plus two equals four on the ground, and demonstrated it with pebbles. Again Tweel caught the idea, and informed me that three plus three equals six. Once more we seemed to be getting somewhere.

"So, knowing that Tweel had at least a grammar school education, I drew a circle for the sun, pointing first at it, and then at the last glow of the sun. Then I sketched in Mercury, and Venus, and Mother Earth, and Mars, and finally, pointing to Mars, I swept my hand around in a sort of inclusive gesture to indicate that Mars was our current environment. I was working up to putting over the idea that my home was on the earth.

"Tweel understood my diagram all right. He poked his beak at it, and with a great deal of trilling and clucking, he added Deimos and Phobos to Mars, and then sketched in the earth's moon!

"Do you see what that proves? It proves that Tweel's race uses telescopes— that they're civilized!"

"Does not!" snapped Harrison. "The moon is visible from here as a fifth magnitude star. They could see its revolution with the naked eye."

"The moon, yes!" said Jarvis. "You've missed my point. Mercury isn't visible! And Tweel knew of Mercury because he placed the Moon at the *third* planet, not the second. If he didn't know Mercury, he'd put the earth second, and Mars third, instead of fourth! See?"

"Humph!" said Harrison.

"Anyway," proceeded Jarvis, "I went on with my lesson. Things were going smoothly, and it looked as if I could put the idea over. I pointed at the earth on my diagram, and then at myself, and then, to clinch it, I pointed to myself and then to the earth itself shining bright green almost at the zenith.

"Tweel set up such an excited clacking that I was certain he understood. He jumped up and down, and suddenly he pointed at himself and then at the sky, and then at himself and at the sky again. He pointed at his middle and then at Arcturus, at his head and then at Spica, at his feet and then at half a dozen stars, while I just gaped at him. Then, all of a sudden, he gave a tremendous leap. Man, what a hop! He shot straight up into the starlight, seventy-five feet if an inch! I saw him silhouetted against the sky, saw him turn and come down at me head first, and land smack on his beak like a javelin! There he stuck square in the center of my sun-circle in the sand—a bull's eye!"

"Nuts!" observed the captain. "Plain nuts!"

"That's what I thought, too! I just stared at him open-mouthed while he pulled his head out of the sand and stood up. Then I figured he'd missed my point, and I went through the whole blamed rigamarole again, and it ended the same way, with Tweel on his nose in the middle of my picture!"

"Maybe it's a religious rite," suggested Harrison.

"Maybe," said Jarvis dubiously. "Well, there we were. We could exchange ideas up to a certain point, and then—blooey! Something in us was different, unrelated; I don't doubt that Tweel thought me just as screwy as I thought him. Our minds simply looked at the world from different viewpoints, and perhaps his viewpoint is as true as ours. But—we couldn't get together, that's all. Yet, in spite of all difficulties, I *liked* Tweel, and I have a queer certainty that he liked me."

"Nuts!" repeated the captain. "Just daffy!"

"Yeah? Wait and see. A couple of times I've thought that perhaps we—" He paused, and then resumed his narrative. "Anyway, I finally gave it up, and got into my thermo-skin to sleep. The fire hadn't kept me any too warm, but that damned sleeping bag did. Got stuffy five minutes after I closed myself in. I opened it a little and bingo! Some eighty-below-zero air hit my nose, and that's when I got this pleasant little frostbite to add to the bump I acquired during the crash of my rocket.

"I don't know what Tweel made of my sleeping. He sat around, but when I woke up, he was gone. I'd just crawled out of my bag, though, when I heard some twittering, and there he came, sailing down from that three-story Thyle cliff to alight on his beak beside me. I pointed to myself and toward the north, and he pointed at himself and toward the south, but when I loaded up and started away, he came along.

"Man, how he traveled! A hundred and fifty feet at a jump, sailing through the air stretched out like a spear, and landing on his beak. He seemed surprised at my plodding, but after a few moments he fell in beside me, only every few minutes he'd go into one of his leaps, and stick his nose into the sand a block ahead of me. Then he'd come shooting back at me; it made me nervous at first to see that beak of his coming at me like a spear, but he always ended in the sand at my side.

"So the two of us plugged along across the Mare Chronium. Same sort of place as this—same crazy plants and same little green biopods growing in the sand, or crawling out of your way. We talked—not that we understood each other, you know, but just for company. I sang songs, and I suspect Tweel did too; at least, some of his trillings and twitterings had a subtle sort of rhythm.

"Then, for variety, Tweel would display his smattering of English words. He'd point to an outcropping and say 'rock,' and point to a pebble and say it again; or he'd touch my arm and say 'Tick,' and then repeat it. He seemed terrifically amused that the same word meant the same thing twice in succession, or that the same word could apply to two different objects. It set me wondering if perhaps his language wasn't like the primitive speech of some earth people—you know, Captain, like the Negritoes, for instance, who haven't any generic words. No word for food or water or man—words for good food and bad food, or rain water and sea water, or strong man and weak man—but no names for general classes. They're too primitive to understand that rain water and sea water are just different aspects of the same thing. But that wasn't the case with Tweel; it was just that we were somehow mysteriously different—our minds were alien to each other. And yet—we *liked* each other!"

"Looney, that's all," remarked Harrison. "That's why you two were so fond of each other."

"Well, I like *you!*" countered Jarvis wickedly. "Anyway," he resumed, "don't get the idea that there was anything screwy about Tweel. In fact, I'm not so sure but that he couldn't teach our highly praised human intelligence a trick or two. Oh, he wasn't an intellectual superman, I guess; but don't overlook the point that he managed to understand a little of my mental workings, and I never even got a glimmering of his."

"Because he didn't have any!" suggested the captain, while Putz and Leroy blinked attentively.

"You can judge of that when I'm through," said Jarvis. "Well, we plugged along across the Mare Chronium all that day, and all the next. Mare Chronium—Sea of Time! Say, I was willing to agree with Schiaparelli's name by the end of that march! Just that grey, endless plain of weird plants, and never a sign of any other life. It was so monotonous that I was even glad to see the desert of Xanthus toward the evening of the second day.

"I was fair worn out, but Tweel seemed as fresh as ever, for all I never saw him drink or eat. I think he could have crossed the Mare Chronium in a couple of hours with those block-long nose dives of his, but he stuck along with me. I offered him some water once or twice; he took the cup from me and sucked the liquid into his beak, and then carefully squirted it all back into the cup and gravely returned it.

"Just as we sighted Xanthus, or the cliffs that bounded it, one of those nasty sand clouds blew along, not as bad as the one we had here, but mean to travel against. I pulled the transparent flap of my thermoskin bag across my face and managed pretty well, and I noticed that Tweel used some feathery appendages growing like a mustache at the base of his beak to cover his nostrils, and some similar fuzz to shield his eyes."

"He is a desert creature!" ejaculated the little biologist, Leroy.

"Huh? Why?"

"He drink no water—he is adapt' for sand storm—"

"Proves nothing! There's not enough water to waste any where on this desiccated pill called Mars. We'd call all of it desert on earth, you know." He paused. "Anyway, after the sand storm blew over, a little wind kept blowing in our faces, not strong enough to stir the sand. But suddenly things came drifting along from the Xanthus cliffs—small, transparent spheres, for all the world like glass tennis balls! But light—they were almost light enough to float even in this thin air—empty, too; at least, I cracked open a couple and nothing came out but a bad smell. I asked Tweel about them, but all he said was 'No, no, no,' which I took to mean that he knew nothing about them. So they went bouncing by like tumbleweeds, or like soap bubbles, and we plugged on toward Xanthus. Tweel pointed at one of the crystal balls once and said 'rock,' but I was too tired to argue with him. Later I discovered what he meant.

"We came to the bottom of the Xanthus cliffs finally, when there wasn't much daylight left. I decided to sleep on the plateau if possible; anything dangerous, I reasoned, would be more likely to prowl through the vegetation of the Mare Chronium than the sand of Xanthus. Not that I'd seen a single sign of menace, except the rope-armed black thing that had trapped Tweel, and apparently that didn't prowl at all, but lured its victims within reach. It couldn't lure me while I slept, especially as Tweel didn't seem to sleep at all, but simply sat patiently around all night. I wondered how the creature had managed to trap Tweel, but there wasn't any way of asking him. I found that out too, later; it's devilish!

"However, we were ambling around the base of the Xanthus barrier looking for an easy spot to climb. At least, I was. Tweel could have leaped it easily, for the cliffs were lower than Thyle—perhaps sixty feet. I found a place and started up, swearing at the water tank strapped to my back—it didn't bother me except when climbing—and suddenly I heard a sound that I thought I recognized!

"You know how deceptive sounds are in this thin air. A shot sounds like the pop of a cork. But this sound was the drone of a rocket, and sure enough, there went our second auxiliary about ten miles to westward, between me and the sunset!"

"Vas me!" said Putz. "I hunt for you."

"Yeah; I knew that, but what good did it do me? I hung on to the cliff and yelled and waved with one hand. Tweel saw it too, and set up a trilling and twittering, leaping to the top of the barrier and then high into the air. And while I watched, the machine droned on into the shadows to the south.

"I scrambled to the top of the cliff. Tweel was still pointing and trilling excitedly, shooting up toward the sky and coming down head-on to stick upside down on his beak in the sand. I pointed toward the south and at myself, and he said, 'Yes—Yes—Yes'; but somehow I gathered that he thought the flying thing was a relative of mine, probably a parent. Perhaps I did his intellect an injustice; I think now that I did.

"I was bitterly disappointed by the failure to attract attention. I pulled out my thermo-skin bag and crawled into it, as the night chill was already apparent. Tweel stuck his beak into the sand and drew up his legs and arms and looked for all the world like one of those leafless shrubs out there. I think he stayed that way all night."

"Protective mimicry!" ejaculated Leroy. "See? He is desert creature!"

"In the morning," resumed Jarvis, "we started off again. We hadn't gone a hundred yards into Xanthus when I saw something queer! This is one thing Putz didn't photograph, I'll wager!

"There was a line of little pyramids—tiny ones, not more than six inches high, stretching across Xanthus as far as I could see! Little buildings made of pygmy bricks, they were, hollow inside and truncated, or at least broken at the top and empty. I pointed at them and said 'What?' to Tweel, but he gave some negative twitters to indicate, I suppose, that he didn't know. So off we went, following the row of pyramids because they ran north, and I was going north.

"Man, we trailed that line for hours! After a while, I noticed another queer thing: they were getting larger. Same number of bricks in each one, but the bricks were larger.

"By noon they were shoulder high. I looked into a couple—all just the same, broken at the top and empty. I examined a brick or two as well; they were silica, and old as creation itself!"

"How you know?" asked Leroy.

"They were weathered—edges rounded. Silica doesn't weather easily even on earth, and in this climate—!"

"How old you think?"

"Fifty thousand—a hundred thousand years. How can I tell? The little ones we saw in the morning were older—perhaps ten times as old. Crumbling. How old would that make *them?* Half a million years? Who knows?" Jarvis paused a moment. "Well," he resumed, "we followed the line. Tweel pointed at them and said 'rock' once or twice, but he'd done that many times before. Besides, he was more or less right about these.

"I tried questioning him. I pointed at a pyramid and asked 'People?' and indicated the two of us. He set up a negative sort of clucking and said, 'No, no, no. No one-one-two. No two-two-four,' meanwhile rubbing his stomach. I just stared at him and he went through the business again. 'No one-one-two. No two-two-four.' I just gaped at him."

"That proves it!" exclaimed Harrison. "Nuts!"

"You think so?" queried Jarvis sardonically. "Well, I figured it out different! 'No one-one-two!' You don't get it, of course, do you?"

"Nope—nor do you!"

"I think I do! Tweel was using the few English words he knew to put over a very complex idea. What, let me ask, does mathematics make you think of?"

"Why—of astronomy. Or—or logic!"

"That's it! 'No one-one-two!' Tweel was telling me that the builders of the pyramids weren't people—or that they weren't intelligent, that they weren't reasoning creatures! Get it?"

"Huh! I'll be damned!"

"You probably will."

"Why," put in Leroy, "he rub his belly?"

"Why? Because, my dear biologist, that's where his brains are! Not in his tiny head—in his middle!"

"*C'est* impossible!"

"Not on Mars, it isn't! This flora and fauna aren't earthly; your biopods prove that!" Jarvis grinned and took up his narrative. "Anyway, we plugged along across Xanthus and in about the middle of the afternoon, something else queer happened. The pyramids ended."

"Ended!"

"Yeah; the queer part was that the last one—and now they were ten-footers—was capped! See? Whatever built it was still inside; we'd trailed 'em from their half-million-year-old origin to the present.

"Tweel and I noticed it about the same time. I yanked out my automatic (I had a clip of Boland explosive bullets in it) and Tweel, quick as a sleight-of-hand trick, snapped a queer little glass revolver out of his bag. It was much like our weapons, except that the grip was larger to accommodate his four-taloned hand. And we held our weapons ready while we sneaked up along the lines of empty pyramids.

"Tweel saw the movement first. The top tiers of bricks were heaving, shaking, and suddenly slid down the sides with a thin crash. And then—something—something was coming out!

"A long, silvery-grey arm appeared, dragging after it an armored body. Armored, I mean, with scales, silver-grey and dull-shining. The arm heaved the body out of the hole; the beast crashed to the sand.

"It was a nondescript creature—body like a big grey cask, arm and a sort of mouth-hole at one end; stiff, pointed tail at the other—and that's all. No other limbs, no eyes, ears, nose—nothing! The thing dragged itself a few yards, inserted its pointed tail in the sand, pushed itself upright, and just sat.

"Tweel and I watched it for ten minutes before it moved. Then, with a creaking and rustling like—oh, like crumpling stiff paper—its arm moved to the mouth-hole and out came a brick! The arm placed the brick carefully on the ground, and the thing was still again.

"Another ten minutes—another brick. Just one of Nature's bricklayers. I was about to slip away and move on when Tweel pointed at the thing and said "rock'! I went 'huh?' and he said it again. Then, to the accompaniment of some of his trilling, he said, 'No—no—,' and gave two or three whistling breaths.

"Well, I got his meaning, for a wonder! I said, 'No breath?' and demonstrated the word. Tweel was ecstatic; he said, 'Yes, yes, yes! No, no, no breet!' Then he gave a leap and sailed out to land on his nose about one pace from the monster!

"I was startled, you can imagine! The arm was going up for a brick, and I expected to see Tweel caught and mangled, but—nothing happened! Tweel pounded on the creature, and the arm took the brick and placed it neatly beside the first. Tweel rapped on its body again, and said 'rock,' and I got up nerve enough to take a look myself.

"Tweel was right again. The creature *was* rock, and it didn't breathe!"

"How you know?" snapped Leroy, his black eyes blazing interest.

"Because I'm a chemist. The beast was made of silica! There must have been pure silicon in the sand, and it lived on that. Get it? We, and Tweel, and those plants out there, and even the biopods are *carbon* life; this thing lived by a different set of chemical reactions. It was silicon life!"

"*La vie silicieuse!*" shouted Leroy. "I have suspect, and now it is proof! I must go see! *Il faut que je—*"

"All right! All right!" said Jarvis. "You can go see. Anyhow, there the thing was, alive and yet not alive, moving every ten minutes, and then only to remove a brick. Those bricks were its waste matter. See, Frenchy? We're carbon, and our waste is carbon dioxide, and this thing is silicon, and *its* waste is silicon dioxide—silica. But silica is a solid, hence the bricks. And it builds itself in, and when it is covered, it moves over to a fresh place to start over. No wonder it creaked! A living creature half a million years old!"

"How you know how old?" Leroy was frantic.

"We trailed its pyramids from the beginning, didn't we? If this weren't the original pyramid builder, the series would have ended somewhere before we found him, wouldn't it?—ended and started over with the small ones. That's simple enough, isn't it?

"But he reproduces, or tries to. Before the third brick came out, there was a little rustle and out popped a whole stream of those little crystal balls. They're his spores, or eggs, or seeds—call 'em what you want. They went bouncing by across Xanthus just as they'd bounced by us back in the Mare Chronium. I've a hunch how they work, too—this is for your informa- tion, Leroy. I think the crystal shell of silica is no more than a protective covering, like an eggshell, and that the active principle is the smell inside. It's some sort of gas that attacks silicon, and if the shell is broken near a supply of that element, some reaction starts that ultimately develops into a beast like that one."

"You should try!" exclaimed the little Frenchman. "We must break one to see!"

"Yeah? Well, I did. I smashed a couple against the sand. Would you like to come back in about ten thousand years to see if I planted some pyramid monsters? You'd most likely be able to tell by that time!" Jarvis paused and drew a deep breath. "Lord! That queer creature! Do you picture it? Blind, deaf, nerveless, brainless—just a mechanism, and yet—immortal! Bound to go on making bricks, building pyramids, as long as silicon and oxygen exist, and even afterwards it'll just stop. It won't be dead. If the accidents of a million years bring it its food again, there it'll be, ready to run again, while brains and civilizations are part of the past. A queer beast—yet I met a stranger one!"

"If you did, it must have been in your dreams!" growled Harrison.

"You're right!" said Jarvis soberly. "In a way, you're right. The dream- beast! That's the best name for it—and it's the most fiendish, terrifying creation one could imagine! More dangerous than a lion, more insidious than a snake!"

"Tell me!" begged Leroy. "I must go see!"

"Not *this* devil!" He paused again. "Well," he resumed, "Tweel and I left the pyramid creature and plowed along through Xanthus. I was tired and a little disheartened by Putz's failure to pick me up, and Tweel's trilling got on my nerves, as did his flying nosedives. So I just strode along without a word, hour after hour across that monotonous desert.

"Toward mid-afternoon we came in sight of a low dark line on the horizon. I knew what it was. It was a canal; I'd crossed it in the rocket and it meant that we were just one-third of the way across Xanthus. Pleasant thought, wasn't it? And still, I was keeping up to schedule.

"We approached the canal slowly; I remembered that this one was bordered by a wide fringe of vegetation and that Mudheap City was on it.

"I was tired, as I said. I kept thinking of a good hot meal, and then from that I jumped to reflections of how nice and home-like even Borneo would seem after this crazy planet, and from that, to thoughts of little old New York, and then to thinking about a girl I know there—Fancy Long. Know her?"

"Vision entertainer," said Harrison. "I've tuned her in. Nice blonde—dances and sings on the *Yerba Mate* hour."

"That's her," said Jarvis ungrammatically. "I know her pretty well—just friends, get me?—though she came down to see us off in the *Ares*. Well, I was thinking about her, feeling pretty lonesome, and all the time we were approaching that line of rubbery plants.

"And then—I said, 'What 'n Hell!' and stared. And there she was—Fancy Long, standing plain as day under one of those crack-brained trees, and smiling and waving just the way I remembered her when we left!"

"Now you're nuts, too!" observed the captain.

"Boy, I almost agreed with you! I stared and pinched myself and closed my eyes and then stared again—and every time, there was Fancy Long smiling and waving! Tweel saw something, too; he was trilling and clucking away, but I scarcely heard him. I was bounding toward her over the sand, too amazed even to ask myself questions.

"I wasn't twenty feet from her when Tweel caught me with one of his flying leaps. He grabbed my arm, yelling, 'No—no—no!' in his squeaky voice. I tried to shake him off—he was as light as if he were built of bamboo—but he dug his claws in and yelled. And finally some sort of sanity returned to me and I stopped less than ten feet from her. There she stood, looking as solid as Putz's head!"

"Vot?" said the engineer.

"She smiled and waved, and waved and smiled, and I stood there dumb as Leroy, while Tweel squeaked and chattered. I *knew* it couldn't be real, yet—there she was!

"Finally I said, 'Fancy! Fancy Long!' She just kept on smiling and waving, but looking as real as if I hadn't left her thirty-seven million miles away.

"Tweel had his glass pistol out, pointing it at her. I grabbed his arm, but he tried to push me away. He pointed at her and said, 'No breet! No breet!' and I understood that he meant that the Fancy Long thing wasn't alive. Man, my head was whirling!

"Still, it gave me the jitters to see him pointing his weapon at her. I don't know why I stood there watching him take careful aim, but I did. Then he squeezed the handle of his weapon; there was a little puff of steam, and Fancy Long was gone! And in her place was one of those writhing, black, rope-armed horrors like the one I'd saved Tweel from!

"The dream-beast! I stood there dizzy, watching it die while Tweel trilled and whistled. Finally he touched my arm, pointed at the twisting thing, and said, 'You one-one-two, he one-one-two.' After he'd repeated it eight or ten times, I got it. Do any of you?"

"*Oui!*" shrilled Leroy. "*Moi—je le comprends!* He mean you think of something, the beast he know, and see it! *Un chien*—a hungry dog, he would see the big bone with meat! Or smell it—not?"

"Right!" said Jarvis. "The dream-beast uses its victim's longings and desires to trap its prey. The bird at nesting season would see its mate, the fox, prowling for its own prey, would see a helpless rabbit!"

"How he do?" queried Leroy.

"How do I know? How does a snake back on earth charm a bird into its very jaws? And aren't there deep-sea fish that lure their victims into their mouths? Lord!" Jarvis shuddered. "Do you see how insidious the monster is? We're warned now—but henceforth we can't trust even our eyes. You might see me—I might see one of you—and back of it may be nothing but another of those black horrors!"

"How'd your friend know?" asked the captain abruptly.

"Tweel? I wonder! Perhaps he was thinking of something that couldn't possibly have interested me, and when I started to run, he realized that I saw something different and was warned. Or perhaps the dream-beast can only project a single vision, and Tweel saw what I saw—or nothing. I couldn't ask him. But it's just another proof that his intelligence is equal to ours or greater."

"He's daffy, I tell you!" said Harrison. "What makes you think his intellect ranks with the human?"

"Plenty of things! First, the pyramid-beast. He hadn't seen one before; he said as much. Yet he recognized it as a dead-alive automaton of silicon."

"He could have heard of it," objected Harrison. "He lives around here, you know."

"Well how about the language? I couldn't pick up a single idea of his and he learned six or seven words of mine. And do you realize what complex ideas he put over with no more than those six or seven words? The pyramid-monster—the dream-beast! In a single phrase he told me that one was a harmless automaton and the other a deadly hypnotist. What about that?"

"Huh!" said the captain.

"*Huh* if you wish! Could you have done it knowing only six words of English? Could you go even further, as Tweel did, and tell me that another creature was of a sort of intelligence so different from ours that understanding was impossible—even more impossible than that between Tweel and me?"

"Eh? What was that?"

"Later. The point I'm making is that Tweel and his race are worthy of our friendship. Somewhere on Mars—and you'll find I'm right—is a civilization and culture equal to ours, and maybe more than equal. And communication is possible between them and us; Tweel proves that. It may take years of patient trial, for their minds are alien, but less alien than the next minds we encountered—if they *are* minds."

"The next ones? What next ones?"

"The people of the mud cities along the canals." Jarvis frowned, then resumed his narrative. "I thought the dream-beast and the silicon-monster were the strangest beings conceivable, but I was wrong. These creatures are still more alien, less understandable than either and far less comprehensible than Tweel, with whom friendship is possible, and even, by patience and concentration, the exchange of ideas.

"Well," he continued, "we left the dream-beast dying, dragging itself back into its hole, and we moved toward the canal. There was a carpet of that queer walking-grass scampering out of our way, and when we reached the bank, there was a yellow trickle of water flowing. The mound city I'd noticed from the rocket was a mile or so to the right and I was curious enough to want to take a look at it.

"It had seemed deserted from my previous glimpse of it, and if any creatures were lurking in it—well, Tweel and I were both armed. And by the way, that crystal weapon of Tweel's was an interesting device; I took a look at it after the dream-beast episode. It fired a little glass splinter, poisoned, I suppose, and I guess it held at least a hundred of 'em to a load. The propellent was steam—just plain steam!"

"Shteam!" echoed Putz. "From vot come, shteam?"

"From water, of course! You could see the water through the transparent handle and about a gill of another liquid, thick and yellowish. When Tweel squeezed the handle—there was no trigger—a drop of water and a drop of the yellow stuff squirted into the firing chamber, and the water vaporized— pop!—like that. It's not so difficult; I think we could develop the same principle. Concentrated sulphuric acid will heat water almost to boiling, and so will quicklime, and there's potassium and sodium—

"Of course, his weapon hadn't the range of mine, but it wasn't so bad in this thin air, and it *did* hold as many shots as a cowboy's gun in a Western movie. It was effective, too, at least against Martian life; I tried it out, aiming at one of the crazy plants, and darned if the plant didn't wither up and fall apart! That's why I think the glass splinters were poisoned.

"Anyway, we trudged along toward the mud-heap city and I began to wonder whether the city builders dug the canals. I pointed to the city and then at the canal, and Tweel said 'No—no—no!' and gestured toward the south. I took it to mean that some other race had created the canal system, perhaps Tweel's people. I don't know; maybe there's still another intelligent race on the planet, or a dozen others. Mars is a queer little world.

"A hundred yards from the city we crossed a sort of road—just a hard-packed mud trail, and then, all of a sudden, along came one of the mound builders!

"Man, talk about fantastic beings! It looked rather like a barrel trotting along on four legs with four other arms or tentacles. It had no head, just body and members and a row of eyes completely around it. The top end of the barrel-body was a diaphragm stretched as tight as a drum head, and that was all. It was pushing a little coppery cart and tore right past us like the proverbial bat out of Hell. It didn't even notice us, although I thought the eyes on my side shifted a little as it passed.

"A moment later another came along, pushing another empty cart. Same thing—it just scooted past us. Well, I wasn't going to be ignored by a bunch of barrels playing train, so when the third one approached, I planted myself in the way—ready to jump, of course, if the thing didn't stop.

"But it did. It stopped and set up a sort of drumming from the diaphragm on top. And I held out both hands and said, 'We are friends!' And what do you suppose the thing did?"

"Said, 'Pleased to meet you,' I'll bet!" suggested Harrison.

"I couldn't have been more surprised if it had! It drummed on its diaphragm, and then suddenly boomed out, 'We are v-r-r-riends!' and gave its pushcart a vicious poke at me! I jumped aside, and away it went while I stared dumbly after it.

"A minute later another one came hurrying along. This one didn't pause, but simply drummed out, 'We are v-r-r-riends!' and scurried by. How did it learn the phrase? Were all of the creatures in some sort of communication with each other? Were they all parts of some central organism? I don't know, though I think Tweel does.

"Anyway, the creatures went sailing past us, every one greeting us with the same statement. It got to be funny; I never thought to find so many friends on this God-forsaken ball! Finally I made a puzzled gesture to Tweel; I guess he understood, for he said, 'One-one-two—yes!—two-two-four—no!' Get it?"

"Sure," said Harrison. "It's a Martian nursery rhyme."

"Yeah! Well, I was getting used to Tweel's symbolism, and I figured it out this way. 'One-one-two—yes!' The creatures were intelligent. 'Two-two-four—no!' Their intelligence was not of our order, but something different and beyond the logic of two and two is four. Maybe I missed his meaning. Perhaps he meant that their minds were of low degree, able to figure out the simple things—'One-one-two—yes!'—but not more difficult things—'Two-two-four—no!' But I think from what we saw later that he meant the other.

"After a few moments, the creatures came rushing back—first one, then another. Their pushcarts were full of stones, sand, chunks of rubbery plants, and such rubbish as that. They droned out their friendly greeting, which didn't really sound so friendly, and dashed on. The third one I assumed to be my first acquaintance and I decided to have another chat with him. I stepped into his path again and waited.

"Up he came, booming out his "We are v-r-r-riends' and stopped. I looked at him; four or five of his eyes looked at me. He tried his password again and gave a shove on his cart, but I stood firm. And then the—the dashed creature reached out one of his arms, and two finger-like nippers tweaked my nose!"

"Haw!" roared Harrison. "Maybe the things have a sense of beauty!"

"Laugh!" grumbled Jarvis. "I'd already had a nasty bump and a mean frostbite on that nose. Anyway, I yelled 'Ouch!' and jumped aside and the creature dashed away; but from then on, their greeting was 'We are v-r-r-riends! Ouch!' Queer beasts!

"Tweel and I followed the road squarely up to the nearest mound. The creatures were coming and going, paying us not the slightest attention, fetching their loads of rubbish. The road simply dived into an opening, and slanted down like an old mine, and in and out darted the barrel-people, greeting us with their eternal phrase.

"I looked in; there was a light somewhere below, and I was curious to see it. It didn't look like a flame or torch, you understand, but more like a civilized light, and I thought that I might get some clue as to the creatures' development. So in I went and Tweel tagged along, not without a few trills and twitters, however.

"The light was curious; it sputtered and flared like an old arc light, but came from a single black rod set in the wall of the corridor. It was electric, beyond doubt. The creatures were fairly civilized, apparently.

"Then I saw another light shining on something that glittered and I went on to look at that, but it was only a heap of shiny sand. I turned toward the entrance to leave, and the Devil take me if it wasn't gone!

"I suppose the corridor had curved, or I'd stepped into a side passage. Anyway, I walked back in that direction I thought we'd come, and all I saw was more dimlit corridor. The place was a labyrinth! There was nothing but twisting passages running every way, lit by occasional lights, and now and then a creature running by, sometimes with a pushcart, sometimes without.

"Well, I wasn't much worried at first. Tweel and I had only come a few steps from the entrance. But every move we made after that seemed to get us in deeper. Finally I tried following one of the creatures with an empty cart, thinking that he'd be going out for his rubbish, but he ran around aimlessly, into one passage and out another. When he started dashing around a pillar like one of these Japanese waltzing mice, I gave up, dumped my water tank on the floor, and sat down.

"Tweel was as lost as I. I pointed up and he said 'No—no—no!' in a sort of helpless trill. And we couldn't get any help from the natives. They paid no attention at all, except to assure us they were friends—ouch!

"Lord! I don't know how many hours or days we wandered around there! I slept twice from sheer exhaustion; Tweel never seemed to need sleep. We tried following only the upward corridors, but they'd run uphill a ways and then curve downwards. The temperature in that damned ant hill was constant; you couldn't tell night from day and after my first sleep I didn't know whether I'd slept one hour or thirteen, so I couldn't tell from my watch whether it was midnight or noon.

"We saw plenty of strange things. There were machines running in some of the corridors, but they didn't seem to be doing anything—just wheels turning. And several times I saw two barrel-beasts with a little one growing between them, joined to both."

"Parthenogenesis!" exulted Leroy. "Parthenogenesis by budding like *les tulipes!*"

"If you say so, Frenchy," agreed Jarvis. "The things never noticed us at all, except, as I say, to greet us with 'We are v-r-r-riends!' Ouch!' They seemed to have no home-life of any sort, but just scurried around with their pushcarts, bringing in rubbish. And finally I discovered what they did with it.

"We'd had a little luck with a corridor, one that slanted upwards for a great distance. I was feeling that we ought to be close to the surface when suddenly the passage debouched into a domed chamber, the only one we'd seen. And man!—I felt like dancing when I saw what looked like daylight through a crevice in the roof.

"There was a—a sort of machine in the chamber, just an enormous wheel that turned slowly, and one of the creatures was in the act of dumping his rubbish below it. The wheel ground it with a crunch—sand, stones, plants, all into powder that sifted away somewhere. While we watched, others filed in, repeating the process, and that seemed to be all. No rhyme nor reason to the whole thing—but that's characteristic of this crazy planet. And there was another fact that's almost too bizarre to believe.

"One of the creatures, having dumped his load, pushed his cart aside with a crash and calmly shoved himself under the wheel! I watched him being crushed, too stupefied to make a sound, and a moment later, another followed him! They were perfectly methodical about it, too; one of the cartless creatures took the abandoned pushcart.

"Tweel didn't seem surprised; I pointed out the next suicide to him, and he just gave the most human-like shrug imaginable, as much as to say, 'What can I do about it?' He must have known more or less about these creatures.

"Then I saw something else. There was something beyond the wheel, something shining on a sort of low pedestal. I walked over; there was a little crystal about the size of an egg, fluorescing to beat Tophet. The light from it stung my hands and face, almost like a static discharge, and then I noticed another funny thing. Remember that wart I had on my left thumb? Look!" Jarvis extended his hand. "It dried up and fell off—just like that! And my abused nose—say, the pain went out of it like magic! The thing had the property of hard ex-rays or gamma radiations, only more so; it destroyed diseased tissue and left healthy tissue unharmed!

"I was thinking what a present *that'd* be to take back to Mother Earth when a lot of racket interrupted. We dashed back to the other side of the wheel in time to see one of the pushcarts ground up. Some suicide had been careless, it seems.

"Then suddenly the creatures were booming and drumming all around us and their noise was decidedly menacing. A crowd of them advanced toward us; we backed out of what I thought was the passage we'd entered by, and they came rumbling after us, some pushing carts and some not. Crazy brutes! There was a whole chorus of 'We are v-r-r-riends! Ouch!' I didn't like the 'ouch'; it was rather suggestive.

"Tweel had his glass gun out and I dumped my water tank for greater freedom and got mine. We backed up the corridor with the barrel-beasts following—about twenty of them. Queer thing—the ones coming in with loaded carts moved past us inches away without a sign.

"Tweel must have noticed that. Suddenly, he snatched out that glowing coal cigar-lighter of his and touched a cart-load of plant limbs. Puff! The whole load was burning—and the crazy beast pushing it went right along without a change of pace! It created some disturbance among our 'v-r-r-riends,' however—and then I noticed the smoke eddying and swirling past us, and sure enough, there was the entrance!

"I grabbed Tweel and out we dashed and after us our twenty pursuers. The daylight felt like Heaven, though I saw at first glance that the sun was all but set, and that was bad, since I couldn't live outside my thermo-skin bag in a Martian night—at least, without a fire.

"And things got worse in a hurry. They cornered us in an angle between two mounds, and there we stood. I hadn't fired nor had Tweel; there wasn't any use in irritating the brutes. They stopped a little distance away and began their booming about friendship and ouches.

"Then things got still worse! A barrel-brute came out with a pushcart and they all grabbed into it and came out with handfuls of foot-long copper darts—sharp-looking ones—and all of a sudden one sailed past my ear—zing! And it was shoot or die then.

"We were doing pretty well for a while. We picked off the ones next to the pushcart and managed to keep the darts at a minimum, but suddenly there was a thunderous booming of 'v-r-r-riends' and 'ouches,' and a whole army of 'em came out of their hole.

"Man! We were through and I knew it! Then I realized that Tweel wasn't. He could have leaped the mound behind us as easily as not. He was staying for me!

"Say, I could have cried if there'd been time! I'd liked Tweel from the first, but whether I'd have had gratitude to do what he was doing—suppose

I *had* saved him from the first dream-beast—he'd done as much for me, hadn't he? I grabbed his arm, and said 'Tweel,' and pointed up, and he understood. He said, 'No—no—no, Tick!' and popped away with his glass pistol.

"What could I do? I'd be a goner anyway when the sun set, but I couldn't explain that to him. I said, 'Thanks, Tweel. You're a man!' and felt that I wasn't paying him a compliment at all. A man! There are mighty few men who'd do that.

"So I went 'bang' with my gun and Tweel went 'puff' with his, and the barrels were throwing darts and getting ready to rush us, and booming about being friends. I had given up hope. Then suddenly an angel dropped right down from Heaven in the shape of Putz, with his under-jets blasting the barrels into very small pieces!

"Wow! I let out a yell and dashed for the rocket; Putz opened the door and in I went, laughing and crying and shouting! It was a moment or so before I remembered Tweel; I looked around in time to see him rising in one of his nosedives over the mound and away.

"I had a devil of a job arguing Putz into following! By the time we got the rocket aloft, darkness was down; you know how it comes here—like turning off a light. We sailed out over the desert and put down once or twice. I yelled 'Tweel!' and yelled it a hundred times, I guess. We couldn't find him; he could travel like the wind and all I got—or else I imagined it—was a faint trilling and twittering drifting out of the south. He'd gone, and damn it! I wish—I wish he hadn't!"

The four men of the *Ares* were silent—even the sardonic Harrison. At last little Leroy broke the stillness.

"I should like to see," he murmured.

"Yeah," said Harrison. "And the wart-cure. Too bad you missed that; it might be the cancer cure they've been hunting for a century and a half."

"Oh, that!" muttered Jarvis gloomily. "That's what started the fight!" He drew a glistening object from his pocket.

"Here it is."

# CHAPTER 3

# SEARCHING THE COSMIC HAYSTACK FOR ETI

*When you flip a coin there is a fifty-fifty chance that it will turn up heads. What are the chances of finding extraterrestrial intelligence if we search the sky diligently?*

*No one can say. We simply do not know enough about the universe to be able to tell. Our ignorance is as wide as the starry galaxies, as deep as the void of space.*

*We do not even know for certain that stars other than our Sun harbor planets where life might take root. Astronomer Bruce Campbell, a leader in the hunt for other solar systems, presents the evidence for the existence of planets orbiting other stars. To date, even though accepted astronomical theory leads to the conclusion that many stars should host planetary systems, no extra-solar planet has been seen. Where observations are lacking, scientists can sometimes use mathematics and statistics to help explore a new area. Frank Drake, one of the fathers of SETI, discusses the equation he devised for the purpose of assessing the chances of finding an extraterrestrial civilization.*

*Your humble editor wishes to make an un-humble declaration here. The first suggestion that organic molecules might exist in space was made by me, in 1962. Since my prediction was published in a science-fiction magazine, hardly anyone paid attention to it. But it was a serious prediction made in a nonfiction article about what would one day come to be called SETI.*

*The following year saw the actual discovery of hydroxyl molecules (OH) in interstellar space, soon followed by discoveries of complex organic molecules. If the chemicals of life exist among the stars, is it likely that they came together only once to form intelligent creatures?*

*Yet there is the possibility that no ETI exists. After all, no one has contacted us yet. David Brin, physicist and award-winning science-fiction author, examines the implications of this interstellar silence.*

*Although radio telescope searches have failed to turn up any intelligent signals from extraterrestrials so far, they have found the chemical constituents of life among the stars. Radio signatures of organic chemicals in interstellar clouds have encouraged the belief that life may be an inherent feature of the Universe and not a fluke occurrence on our one little lonely world.*

# A PLACE TO LIVE: EXTRASOLAR PLANETS

## BY BRUCE CAMPBELL

For life to originate, thrive, and evolve to intelligent forms at many locations throughout our galaxy, there must almost certainly be an abundance of extrasolar planets—planets similar to Earth orbiting other stars. Science-fiction writers have suggested that exotic life forms could develop in less hospitable environs, such as on the surface of a neutron star (the extremely dense remnant left after a supernova explosion), or in an interstellar dust cloud. However, Earth-like planets have all the essential ingredients without stretching the probabilities: (1) a stable source of energy (the parent star), which is far enough away to allow molecules to form, but not much farther; and (2) plenty of raw materials in readily usable forms.

With the crucial role that extrasolar planets play in our search for extraterrestrial life, it is disappointing to learn that we do not even know if such planets exist. Of course astronomers expect that they are out there, but despite numerous claims, there is as yet *no hard evidence for the existence of even one planet outside our solar system.* This may seem surprising, given the tremendous advances in sensitivity of astronomical techniques, and the proliferation of ideas on how to find these objects.

### CIRCUMSTELLAR DISKS: A CLUE TO PLANET FORMATION

It is now widely accepted by astronomers that the planets in the solar system were formed at about the same time as the Sun itself. They were produced in a flattened disk of material left over after the collapse of a slowly rotating interstellar cloud of gas and dust. Once the disk had settled to form a thin sheet, the particles of dust began to clump together to form solid bodies called planetesimals. These in turn slowly aggregated to form larger objects that were ultimately able to sweep up with their gravitational pull all the remaining material in their orbital zone. While many details of planetary formation remain obscure, the general picture of the growth of planets in a disk is almost certainly correct. There are no remotely plausible alternatives.

It is therefore encouraging that astronomers have found around young stars disks that may be in the process of forming into planets. Of particular interest are the class of stars (named after a faint example in the constellation of Taurus) called the T Tauri stars. These stars have formed within the last few million years. Their relative youth (the Sun is five *billion* years old) is inferred from their location—they are found within or near dense interstellar clouds from which they presumably condensed. As well, they have high rates of rotation and high levels of magnetic activity, characteristics of young stars. This activity creates large numbers of starspots (analogous to sunspots) on young stars, the stellar equivalent of adolescent acne!

Several lines of evidence show that T Tauri stars have disks of dust and gas around them that would provide ample material from which to generate a planetary system. In particular, observations of such stars in the infrared (heat radiation) show that a great deal of energy is coming from them at these wavelengths, much more than from older stars. Evidently the infrared radiation is coming from the dust particles in the disks, which are glowing at these wavelengths because they are being heated by shorter wavelength starlight. Since older stars do not produce large amounts of infrared emission, we must infer that the material comprising the disks goes *somewhere* as stars pass into middle age. Perhaps the disks disappear by changing into planets. However, this is by no means proven, as the disks could just as well be blown away by stellar winds, which are streams of subatomic particles ejected from stars at high velocities.

An indication that stellar winds do not clear away *all* of the disk material comes from recent discoveries of very tenuous disks around a few dozen nearby old stars. These are quite different in character from the T Tauri disks, as the old stellar disks seem to be made of relatively large particles, and contain only a tiny fraction of the material found in the disks of young stars. The disks around old stars are so faint and difficult to detect that they can be seen only around the very nearest stars, but it looks as though they may be present around virtually every star. Figure 1 shows an example of one such disk around the A-type star Beta Pictoris.

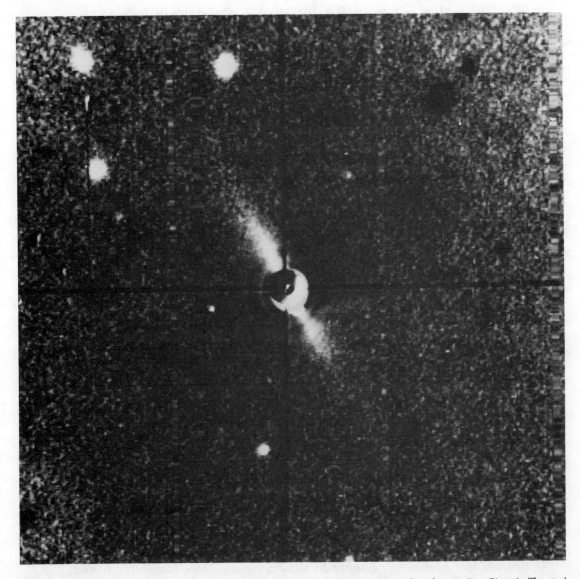

**Figure 1.** This computer-processed image shows the faint disk, seen edge-on, surrounding the star Beta Pictoris. The star's image was cancelled out to allow the faint disk to be seen. The black cross is a computer artifact. *(Photo: Courtesy NASA.)*

These tenuous disks themselves may not have any direct relationship with planetary systems. However, the discovery that some of these disks have a "hole," or empty region in the vicinity of their parent star, has generated a great deal of interest, because the hole could be telling us that planets are present. The idea is that such a hole is not expected to remain empty, unless something is continually sweeping up the particles that drift in from the outer regions of the disk. Perhaps a system of planets is collecting the material from the hole. That the holes are roughly comparable in size to the solar system is consistent with this picture. It would be unwise, however, to take this suggestion as fact at the present. There are still too many uncertainties in both the observations and the theoretical interpretation to conclude that the few known cases of holes in circumstellar disks necessarily imply the presence of planets. All we can say for the moment is that this is a strong hint.

Another such hint had once been taken from the study of rotation rates for stars. It was noticed that stars much more massive than the Sun generally rotate rapidly, often as much as one hundred times faster, whereas stars similar to the Sun usually spin at a leisurely pace. While the Sun does not have much angular momentum (tendency to keep spinning), the planets in the solar system do, by virtue of their large orbits. If the planets were drawn into the Sun, they would cause it to rotate as fast as a massive star, in the same way as a figure skater spins faster if she pulls her arms into her body. With the assumption that all proto-stellar clouds starts out with about the same angular momentum, the hypothesis was advanced that stars like the Sun rotate slowly because they have planetary systems to carry the extra angular momentum. This led to the notion that virtually all slowly rotating stars—that is, the majority of stars in our galaxy—have planets in orbit about them.

Unfortunately we now know that this scenario is probably wrong. Some T Tauri stars are similar in mass to the Sun, and yet rotate at high rates comparable to massive stars. We now understand that Sun-like stars gradually slow down their rate of spin as they age, through the interaction of stellar winds with the star's magnetic field. It is the stellar winds themselves that act like brakes, continuing to slow a rotating star long after planets might have formed.

So, while they are not ruled out, planets are not needed after all to explain the slow rotation of stars like the Sun. This goes to show that one should beware of "hints" to the existence of other planetary systems!

## FAILED STARS: THE BROWN DWARFS

Astrophysicists believe that nuclear fusion, which is the energy source for normal stars, will not commence in bodies less than 8 percent the mass of the Sun. (Jupiter, the largest planet, is only 0.1 percent the Sun's mass, or eighty times smaller than this limit.) Since these substellar objects, called "brown dwarfs," do not generate any readily detectable energy, astronomers have faced the same dilemma in searching for them as they have with planets: how does one "see" something that produces virtually no light? It must be emphasized that brown dwarfs are not the same as planets. Brown dwarfs are really a variation of normal stars, since, like stars, they form from a collapsing cloud of gas and dust, rather than by aggregation of material in a circumstellar disk. This means that brown dwarfs can exist in isolation, separate from any normal star, which is not the case for planets.

Counts of ordinary stars indicate that there are many more stars of low mass than high mass per unit volume of space. (A schematic representation of stellar populations by mass is shown in Figure 2.) Since brown dwarfs are so difficult to detect, it was, until recently, not clear if this trend for stars continues upward at the low mass end, in which case substellar objects would abound, or if the trend reverses itself. Simple extrapolation of the trend for normal stars would suggest that brown dwarfs are indeed numerous, so numerous in fact that they could outnumber all of the stars in our galaxy.

Another piece to the puzzle comes from the so-called "missing mass" problem. From studies of the motions of stars in the solar neighborhood we know that as much as half of the mass of our galaxy is in a form as yet unidentified. Putting this together with the extrapolation of the mass-frequency trend for normal stars led to the suspicion that brown dwarfs are indeed abundant, to the point where they account for as much of the mass in our galaxy as all of the stars and interstellar clouds combined. Brown dwarfs are not the only possible explanation for the missing mass, but they have headed the list of suspects for some time.

Given this background, it is no wonder that astronomers have been surprised to find that brown dwarfs are *not* numerous after all. Sensitive surveys for brown dwarfs conducted in the last few years should have easily uncovered some of these elusive objects if they were so abundant as to account for the missing mass. Yet there are at present just a few proposed candidates for brown dwarfs, and even these cases are in doubt. It is difficult to imagine how a large population of these objects could still be hiding out somewhere. The conclusion is that brown dwarfs are not numerous; they are rare. Furthermore, it is unlikely that brown dwarfs make up the missing mass. To explain that problem we must resort to some other alternative, such as exotic subatomic particles, or perhaps black holes.

One might at first think that the absence of brown dwarfs does not augur well for the existence of planets, which have even lower masses. There is really no reason to be pessimistic, however. It must be remembered that even though brown dwarfs and planets may have similarly small masses, they really are quite distinct in their origins, a point sometimes not appreciated even by professional astronomers.

In fact, the absence of brown dwarfs actually enhances our prospects for finding planetary systems. Since no brown dwarfs have been found as close companions to normal stars, these stars could harbor planets in stable orbits. If brown dwarfs had turned up close to Sun-like stars, and if planets could form near the star in such systems, then they would be quickly thrown off into deep space by the gravitational forces of the star–brown dwarf pair. The absence of brown dwarfs also removes the ambiguity in what to call very-low-mass objects—anything on the order of the mass of Jupiter would have to be classified as a planet.

## MINUTIAE IN THE COSMIC HAYSTACK: THE DILEMMA OF DIRECT DETECTION

One of the hardest things to appreciate when considering the search for other planetary systems is just how small and insignificant planets really are. Earth is not the smallest planet in the solar system, but it certainly is not the largest. Jupiter is by far the most massive, weighing in at more than three hundred times the mass of Earth. On the scale of the stars, however, even Jupiter is insignificant: the Sun is more than *one thousand times the mass of Jupiter.*

It is not just the huge difference in mass that matters. To see an object in space, we must detect light coming from it. Stars produce ample quantities of light from nuclear reactions deep in their interiors, but planets produce essentially no energy of their own, and are visible only through the tiny amount of starlight that they reflect. The problem, then, comes from the fact that the Sun produces about one billion times as much light as is reflected by Jupiter! Viewed from the nearest star, Proxima Centauri, which is 4.2 light years or 25 trillion miles away, Jupiter would appear to be very

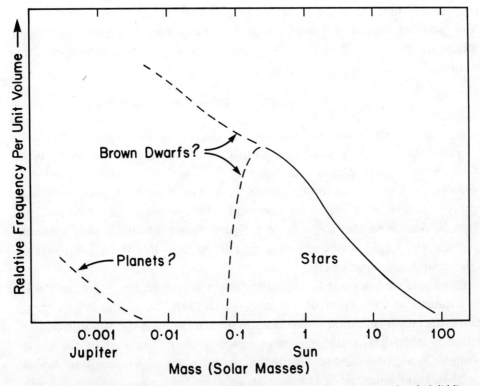

**Figure 2.** The relative populations per cubic light year of stars, planets, and brown dwarfs. Solid lines are based on observation, while dashed lines represent extrapolations. The brown dwarf population might be numerous (rising curve) or rare (falling curve). *(Diagram supplied by Bruce Campbell.)*

close to its Sun, and would be completely lost in the glare of sunlight. It would be like trying to see a firefly's glow in the beam of a spotlight viewed at a distance of several thousand miles.

This is not to say that it is totally impossible to directly see planets orbiting other stars. Astronomers are considering ways of surmounting the difficulties, but some substantial advances in technology are needed to achieve this. Direct detection will require a specialized telescope and imaging system designed to greatly reduce the starlight normally scattered and diffracted by telescope optics. This equipment must be put into space to eliminate the blurring effects of Earth's atmosphere. At least one such scheme is under study by scientists at the Jet Propulsion Laboratory and Perkin-Elmer Corporation. Despite some promising developments in this area, it is likely that direct detection is still well in the future.

## SIGNS OF PLANETS IN STELLAR WOBBLES

Fortunately, there are alternate ways to detect extrasolar planets besides "seeing" them directly. Even though planets are comparatively small, they do influence their parent stars slightly through their mass and orbital motion. For example, as Jupiter moves around in its orbit, the Sun executes a tiny "counter-orbit" about the common center-of-mass of the two bodies. This is similar to the rotation of a figure skater spinning with a small partner. The larger skater moves in a small circle, while the smaller skater goes round in a large circle.

Of course we normally think of the Sun as immobile, fixed like the Rock of Gibraltar at the center of the solar system, but this is not strictly true. Since the Sun is one thousand times the mass of Jupiter, the Sun moves in a counter-orbit that is one thousand times smaller than that of the giant planet. On astronomical scales the Sun's orbit is tiny; its radius amounts to only 460,000 miles, or just slightly larger than the radius of the Sun itself. While the Sun actually moves under the influence of all the planets, by far the largest effect is that due to Jupiter. If we could detect comparable motion in another star, then we could infer the presence of a planet that we cannot at present see directly.

Finding such a motion in another star is a difficult task, however. One way to look for it would be to measure accurately the change in position of a target star over time, to see if on top of its general motion across the sky there is a wobbly or wavy motion that would signal the presence of a dark companion. The star's position has to be measured relative to a set of distant stars seen in the same general direction; such observations are made routinely by astronomers for a variety of purposes. Unfortunately, these measurements are not normally accurate enough to detect the influence of a planet. The angular motion of the Sun in its counter-orbit with Jupiter, when viewed from Proxima Centauri, would be equivalent to the angle that a dime makes when viewed from a distance of three hundred miles. Stellar position measurements normally fall short of this level of accuracy by about a factor of three. Hence some fairly modest improvements to existing techniques would enable astronomers to detect the wobbly motions in stars due to planets.

In recent decades, a few astronomers have thought that their measurements were sufficiently accurate, and have reported success in finding planets. Such reports have tended to be sensationalized, and this, in turn, has led to the widespread notion that extrasolar planets have already been discovered. Consistently such reports have turned out to be wrong, or at least have remained unconfirmed. What is worse, new information contradicting the old often does not reach the public.

The most notorious such case is that of Barnard's star, which at six light-years is one of the nearest stars to the Sun. In 1963, Peter van de Kamp, an astronomer at Sproul Observatory, reported that extensive study of Barnard's star revealed a tiny wobble superimposed on its space motion. The magnitude of this wobble was said to be consistent with a Jupiter-sized planet, although a later analysis suggested the presence of *two* planet-sized bodies. Naturally this report attracted a great deal of attention at the time, while subsequent analyses of the same and other data went virtually unnoticed.

One follow-up study looked at comparable observations from the Allegheny and Van Vleck observatories. These showed no sign of the wobble reported by van de Kamp. (See Figure 3 for a comparison of the two sets of data.) Yet another study, a fresh look at the original Sproul Observatory measurements, suggested that mechanical adjustments to the telescope over the years had introduced subtle shifts, which could have resulted in the spurious appearance of a wobble. With these contradictory reports, the case for a planet orbiting Barnard's star remains, in the opinion of almost all astronomers, far from proven.

The principal conclusion that emerges from such cases is that higher-accuracy measurements are required to be sure that a planet-induced wobble is present. One group of astronomers, led by George Gatewood of the Allegheny Observatory, has developed a very promising method of improving on older techniques; it can measure stellar positions at least twenty times more accurately than was previously possible. The Allegheny group has begun a project to monitor a number of nearby stars with the objective of eventually detecting the motion induced by Jupiter-like planets. Given that the orbital period of Jupiter is nearly twelve years, no one expects dramatic results from this project very soon.

## THE PROMISE OF THE DOPPLER TECHNIQUE

Another way to look for the effects of a planet is to measure the motion of a star along the line-of-sight, toward or away from us, using what is known as the Doppler effect. This effect causes the frequencies of light coming from a star to change, shifting to higher frequency (bluer light) if the star is approaching us, and to lower frequency (redder light) if it is receding.

An analogous phenomenon occurs when you hear the whistle of a fast train. The whistle sounds as if it has a higher pitch (i.e., higher frequency) as the train approaches, and lower pitch when it has passed and is receding. By carefully measuring the frequency of the sound waves from such a whistle, one could determine if the train is approaching or receding, and at what rate of speed.

Similarly, by observing a spectrum of starlight, an astronomer can tell how fast a star, which is a source of light waves, is moving toward or away from the Earth. Astronomers measure the frequencies of certain features in the spectrum, where particular colors of starlight have been filtered out by atoms in the atmosphere of the star. Laboratory studies of these atomic species yield the frequences of light corresponding to these spectral features for a source at rest. Measurement of the frequencies in the starlight will show by how much they are higher or lower than normal, which in turn reveals how fast the star is approaching or receding.

If the star happens to have an orbiting companion, then the velocity of the star will appear to change as it goes around in its counter-orbit. In fact, there would be an oscillatory motion, toward and away, superimposed on the general motion of the star through space. Such cyclic motion is routinely found by astronomers in double stars, and from studies of these systems we derive most of our knowledge about the masses of stars. In such cases the orbital velocity of a star is typically in the range of 20,000 to as much as 200,000 miles per hours. If, however, the companion is a planet,

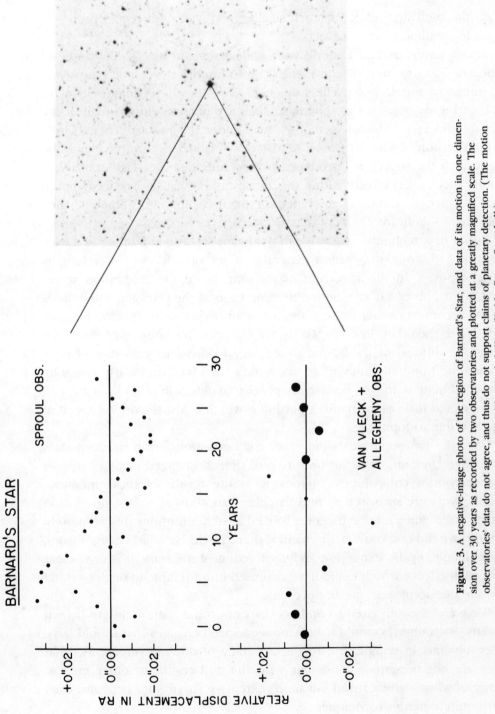

**Figure 3.** A negative-image photo of the region of Barnard's Star, and data of its motion in one dimension over 30 years as recorded by two observatories and plotted at a greatly magnified scale. The observatories' data do not agree, and thus do not support claims of planetary detection. (The motion shown is in Right Ascension, or celestial longitude.) *(Supplied by Bruce Campbell.)*

then the oscillatory effect is much smaller, and was, until recently, considered to be impossible to detect.

About ten years ago I, along with colleagues Gordon Walker and Stephenson Yang of the University of British Columbia, developed a new way to measure the Doppler effect in stars. This novel technique involves calibrating wavelengths in spectra in such a way as to eliminate all of the instabilities of traditional methods. These older techniques generally rely on calibration of wavelengths with a standard laboratory source. Since the light from the source is introduced before and after the star is observed, subtle mechanical effects within the apparatus can cause displacements in the wavelength direction unrelated to the Doppler effect; that is, they introduce significant errors of measurement.

In our new technique, we place in the beam of starlight a tube filled with hydrogen fluoride gas, which generates a set of reference wavelengths superimposed on the spectrum of the star. Since the calibration wavelengths are observed at exactly the same time as the starlight, instabilities in the apparatus cannot cause a displacement of one set of spectral features relative to the other. In other words, we eliminate the sources of error that plague traditional methods, and so are able to measure velocities of stars about one hundred times more accurately than was previously possible. The technique is, in fact, sufficiently precise to detect in other stars a speed change of as little as thirty miles per hour, which is just the orbital speed of the Sun due to Jupiter.

Over the past eight years our team has been monitoring eighteen stars similar to the Sun, with the objective of detecting speed changes due to planetary-mass companions. In looking at the results of these measurements, we were surprised to find that the stars showed almost no motion whatsoever. Since no one had ever looked at stellar motions this accurately before, we did not know beforehand if there might be rapid changes from, for example, cyclic expansion and contraction of the stars. It is now clear that such effects, which could have confused our attempts to detect orbital motion due to planets, are not present.

We were also surprised to find no signs of orbital motion due to brown dwarfs. Such objects would have produced speed changes that should have been obvious in comparison to the accuracy of our measurements. That they are not present is consistent with the null results of other groups, many of whom have carried out more extensive surveys for brown dwarfs with quite different techniques.

Our most exciting result, however, is that we have found gradual motion

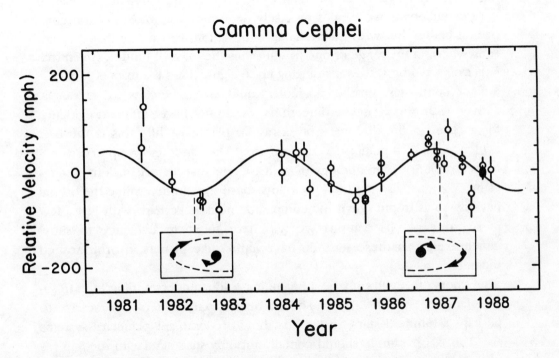

**Figure** 4. The cycle in velocity of approach and recession of the star Gamma Cephei is good evidence of a planet with a three-year period. Note: the detected velocity is only about 50 miles per hour. *(Supplied by Bruce Campbell.)*

that could be due to planetary companions. And what is more, we have found significant motion in *nine* of the eighteen stars we have been following. That we find motion in so many stars is encouraging, especially when we consider that previous claims of planetary companions have generally involved only one or two stars.

In most of these nine cases we have not yet seen a full orbital cycle. This is not surprising, if the companion objects have periods of revolution similar to the twelve years of Jupiter. (For one star, Gamma Cephei, we have actually observed two orbital cycles. See Figure 4.) With the orbital periods unknown, we cannot precisely determine the masses of the companion bodies, but with the available information we can say that they are somewhere in the range of one to ten times the mass of Jupiter. Given that larger objects—brown dwarfs ranging up to eighty times the mass of Jupiter— are not found, these small companions would have to be classified as planets.

Even with nine tentative detections, we do not have sufficient evidence for a claim to the discovery of extrasolar planets. While this is the best evidence for such planets to date, we would like another group of astronomers to confirm our findings in at least one case before making such a claim. We must also continue to follow these stars to determine the orbital periods, so as to pin down the companion masses. Perhaps with just a few more years of observations we will have sufficient evidence to show conclusively that these stars do have Jupiter-like planets in orbit around them.

Of course this will only be a beginning to our understanding of extrasolar planetary systems. There should be an extension of our search to perhaps a hundred stars, to test just how predominant planetary systems are. This larger sample should contain not only stars similar to the Sun (as are the majority in our present program), but very different ones as well. Do massive, hot stars, or the giant stars that have nearly ended their energy-producing life have planets? What about double star systems—can planets form in them, and can they survive for billions of years in stable orbits? Have planets condensed around the oldest stars in the Milky Way, which have a much lower content of the heavy elements from which rocky planets like Earth form? Our anthropomorphic view says that smaller planets will be the focus of our search for life elsewhere, and only with ever more sensitive techniques can we hope to detect them. Devising such ultrasensitive methods will be a major challenge for the next generation of astronomers.

## THE PROSPECTS FOR DETECTING EXTRATERRESTRIAL LIFE

Given that we have just a few clues to the existence of other planetary systems, how can we possibly address the question of whether there is life on these planets? While we cannot directly confront this problem at present, we can avoid the issue of whether such life has a planetary home if we only consider looking for intelligent life. Then we can contemplate communicating with an advanced civilization by means of, for example, radio signals. It is conceivable that we could do this, and efforts in this direction are discussed in other chapters of this book.

But are there any civilizations out there sending signals our way? Of course we will not know until we actually search for these signals, but it is interesting to reflect on the factors that would determine the number, $N$, of civilizations in our galaxy willing and able to communicate with us. Dr. Frank Drake, an astronomer at the University of California at Santa Cruz, is credited with first devising a simple equation for this number:

$$N = R_* f_p n_e f_l f_i f_c L$$

The factors in the Drake Equation are:

$R_*$ = the number of *new stars formed* in our galaxy each year

$f_p$ = the fraction of those stars that have *planetary systems*

$n_e$ = the average number of planets in each such system that can *support life*

$f_l$ = the fraction of such planets on which *life actually originates*

$f_i$ = the fraction of life-sustaining planets on which *intelligent life* evolves

$f_c$ = the fraction of intelligent life-bearing planets on which the intelligent beings develop the means and the will to *communicate* over interstellar distances

$L$ = the average *lifetime* of such technological civilizations

As one might well imagine, the best that we can do for most of the factors in this equation is *guess.* For $R_*$, astronomers do know that the stellar birthrate averaged over the lifetime of the Galaxy is on the order of ten stars per year. This could be an overestimate, however, as the present rate of star formation is probably lower than it was in the past. Also, we should exclude certain classes of stars that are unlikely to have life-sustaining planets, such as very close double stars, stars with short life spans, and stars in the densely packed central regions of our galaxy. Based on the results to date of our project to measure stellar velocities accurately, where half the stars monitored show velocity changes apparently due to small companions, one might conclude that $f_p$ (stars with planetary systems) is roughly 0.5. Obviously this estimate is highly uncertain.

While there is considerable uncertainty in these estimates of $R_*$ and $f_p$, this is a minor detail in comparison to how little we know about the other factors. For $n_e$, $f_l$, $f_i$, and $f_c$, we face the problem that we do not know very much about how life originated on Earth, let alone how it might develop elsewhere. We know very little about what factors are critical in making a planet habitable, and we know even less about how life evolves on average once it appears. There is not even a concensus among biologists that evolution to intelligence is inevitable. Perhaps the norm is single-celled organisms, which reigned as the sole inhabitants of Earth for more than half of the time that life has existed here.

Of course the greatest uncertainty is in estimating $L$, the average lifetime of technological civilizations. Do most advanced societies destroy themselves, or meet with some natural catastrophe, before they can develop the means for interstellar communication? Or do they avert disaster and achieve a stable existence for millions or billions of years? There does not seem to be any way we can answer this question, unless we ask the extraterrestrials. Meanwhile, for $L$ we have only the example of our own civilization to go by, and, paradoxically, only after we have disappeared would we have the answer to the question!

So the Drake Equation really does not tell us much, since even reasonable guesses for the various factors lead to $N$ in the range of 1 (i.e., we are alone) to 1,000,000. This exercise tells us how much more we need to know.

What, then, are the prospects for finding intelligent extraterrestrial life? While we cannot answer this question, we should not be discouraged from looking. Despite the difficulties, there is no doubt of the profound impact of a successful discovery on how we view ourselves and how we view the universe around us. We should not be under any illusion that this task is going to be easy. There are so many places and ways to look for artificial signals, which are likely to be both weak *and* confused with ones which we generate ourselves, that with an ambitious search it still could take decades before contact is made, even if such civilizations are scattered across the Galaxy. To make matters worse, this approach *assumes* that these civilizations are transmitting signals to begin with, which for various reasons they might choose not to do. So the search for extraterrestrial intelligence—SETI—should be pursued, but the enterprise must be considered high-risk. This risk is balanced by the potential for a very high return: knowledge of, and possibly communication with, other intelligent beings.

There is, however, another, less glamorous way to determine if life exists elsewhere in the Galaxy. This would require a systematic, step-by-step approach, with several small advances possibly leading to the finding of life, which need not be intelligent. These steps would go as follows:

First, we would have to find extrasolar planetary systems. Obviously the large planets like Jupiter, will turn up initially. We are well on the way to achieving this goal with our accurate stellar-velocities project, and in a few years we should be able to tell if such large planets are common, or perhaps rare.

Next, with highly sensitive direct or indirect techniques, we could detect planets much smaller than Jupiter. Again we would need to know whether or not these are common. Also, we would want to know how frequently they have sufficient mass to retain a dense atmosphere, and how often they are in the thermally habitable zone of the parent star. (As the astronomer Tobias Owen once said, the situation in our solar system could be described as Goldilocks and the three planets—Venus is too hot, Mars is too cold, and Earth is just right.)

Ultimately, we could determine if life is present on such planets by, for example, searching for the signature of oxygen in spectra of their reflected starlight. Oxygen is plentiful in Earth's atmosphere only because of respiration by plants, and so spectral features of oxygen are a clear signpost to distant observers that life is present on Earth. The detection of oxygen in the atmosphere of another planet would demonstrate that at least photosynthesizing life is present.

While each of these steps represents a significant technological challenge, there is no fundamental barrier to success; we can at least conceive of ways to make the requisite measurements. Of course "success" here does not necessarily mean detecting evidence of primitive (plant) life—

perhaps life similar to that on Earth is rare, or habitable planets are uncommon. But in such a circumstance we could still be successful in answering the question of how pervasive photosynthesizing life is.

The same cannot be said for lack of success from a thorough SETI program, since the absence of signals could mean either that advanced civilizations are not present *or* that they are not transmitting. (There are parallels here on Earth—even if the Europeans of the fifteenth century had thought of looking, could they have ever detected the Mayan culture without going to Central America?) So technological civilizations may well be out there, but we would not know it. Still, success in SETI would have such a profound impact that we can easily justify devoting time and money to the effort.

A step-by-step approach to finding life, starting with the detection of extrasolar planets, can also be easily justified. While it yields a somewhat lower return than SETI—if finding only primitive life can be called that—it also has lower risk in that "success" (showing if photosynthesizing life is present) is virtually assured. The conclusion, then, is that we should pursue both approaches with all available resources. And we should take heart that new developments, both in SETI techniques and in methods for finding extrasolar planets, show great promise. We could well be on the threshold of some very exciting discoveries in these areas.

# THE DRAKE EQUATION: A REAPPRAISAL

## BY FRANK D. DRAKE

It has been twenty-eight years since I first wrote the formula that has come to be known as the Drake Equation. The motivation was not to enlighten the world, but rather to organize rationally the program for the first scientific meeting ever, as far as I know, on the subject of extraterrestrial intelligent life. I was the lone organizer for that meeting, which would be attended by almost everyone in the world active in the subject—all twelve of them. Looking back over all those years, I think the equation has survived the test of time quite well, a rarity in science, and has become the basis of major searches for life now underway.

The survival of the equation is due, I think, to the fact it was constructed in the best tradition of science: The preferred explanation of a phenomenon is the simplest explanation. Don't get "clever" until the facts make you do so. Our knowledge then and now of the evolution of stars, planets, the molecules of life, and of systems of living things said that our star, its planets, and life on Earth, in all their complexity, are the result of completely normal processes that occur readily in this Universe. No freak events are required. Thus the basic premise behind the equation is that what happened here will happen with a large fraction of the stars as they are created, one after another, in the Milky Way Galaxy and other galaxies. People unfamiliar with the accepted pictures of cosmic and biological evolution might think the equation is highly speculative; in fact, it is just the opposite, since the phenomena it assumes to take place in the Universe are only those we are sure have taken place at least once.

There is actually an exception in the equation to that last statement, and one that causes the result to be very conservative and to bend over backward not to overestimate the abundance of detectable civilizations in the Universe. This is the inclusion of the factor $L$, the mean longevity of civilizations in a detectable state. We actually do not know that there is a limitation to this longevity; only the study of other civilizations will reveal the truth. However, the assumption of a limited longevity introduces into the equation our one area of almost total ignorance in such a way that it minimizes our estimates of the numbers of civilizations we might detect. In the end, then, the equation is carefully not speculative. Those who use it to contemplate life in the Universe, or to plan searches, are going to great lengths to be nonspeculative fuddy-duddies.

Twenty-eight years ago the typical number of technological civilizations given by the equation for the Milky Way was roughly ten thousand. This came from estimates based on observations and theory, in which all the factors in the equation but the last multiplied out to a value of about one, and we guessed that $L$ was of the order of ten thousand. The observational evidence providing actual values for the factors in the equation has greatly improved over the years, but the end result has not changed significantly. There have been a multitude of laboratory experiments tracing the possible chemical pathways by which life might develop. We have seen in the interstellar clouds, in the comets, and in the meteorites the molecules of life or their chemical precursors, confirming that the chemistry of life is ubiquitous, and, quite separately, we have seen that the march toward life on Earth might have started even before our planet was formed. Similarly, in the fossil record we have seen evidence that there is more than one path to an intelligence species. Recent fossil evidence shows that there were a few relatively large-brained dinosaurs, which were well along the path to intelligent status; it is only through chance that we are the first on Earth.

In the same vein, our exploration of the planets has shown that there is a bewildering variety of planets and satellites just in our own system; for example, in the Jovian system we have Europa, with its surface of ice covering, probably, a complicated ocean, and right next to it the bone-dry Io, rich with active volcanos of sulphur. In recent times there has been a wealth of new information indicating that a large fraction of other stars has planetary systems. The underlying truth behind these discoveries, and others, such as that of the pulsars and quasars, is not surprising: In the Universe, anything that can happen will happen. And often. This is perhaps the Murphy's Law of the Universe. The existence of life, and intelligent, technology-exploiting life, in large quantities, should not be a surprise.

Over all these years we have surmised that the most promising way to search was with radio telescopes. It is at wavelengths of a few centimeters that the Universe is darkest and quietest; we know this well, and so will the others. It is at such wavelengths that we can detect the faintest signals with the least expenditure of resources, and also where another civilization can communicate with us at least cost, if it so desires. Until recently, nevertheless, it was hard for us to search, because we didn't have the ability to listen to many frequency channels at once. Modern computer technology has changed all that, and now we can listen to millions of channels simultaneously.

The equation tells us that, with that capability, it is worth searching. There is a real chance that with realistic resources, and in an acceptable time, we will find one of those other civilizations. Wisely, we are exploiting this opportunity through the several searches now planned or underway. We have entered the age of the greatest of all explorations, the search for intellect throughout the Universe.

# MYSTERY OF THE GREAT SILENCE

## BY DAVID BRIN

I first heard the subject of extraterrestrial intelligence brought up at a scientific seminar in 1968, when a speaker suggested that pulsars— then newly discovered—might be beacons of an advanced civilization. He was only partly serious, but it was soon clear that most of those with tenure didn't like this kind of talk at all. Only a few years later, however, some of those who were angriest in 1968 applauded when Carl Sagan unveiled at Caltech a gold "message" plaque to be placed upon Pioneer 10, the first human artifact that would leave the solar system.

Soon scientists were discussing not whether extraterrestrial intelligences exist, but how to go about listening for signals from our nearest neighbors, adapting radio telescopes for the search and asking, "Is anyone out there?"

But the voices of the critics never really went away. For no sooner had it become legitimate to inquire, "Where are they?" than new questions were raised asking, "Why aren't they here already?"

But let's not get ahead of ourselves. To early xenologists it was danger- ous enough talking about alien life forms without risking one's scientific reputation talking about interstellar travel. Thus, modern scientific xenology first dealt with the possibility of life springing up in isolation among the stars—yielding islands of sentience separated by vast distances and for all time.

Early students of this new science, no matter how daring, were faced with one major limitation: a near total lack of data. Still, certain scientific discoveries, combined with a useful philosophical tool, gave researchers the courage to make crude estimates.

First, it was found almost ridiculously easy to make amino acids and other precursors to living matter from abundant molecules such as methane, ammonia, and cyanogen. Harold Urey and Stanley Miller subjected a solution of these substances to electrical discharge and ultraviolet radiation and got an organic "soup" in short order. Furthermore, during the last two decades, radio astronomers have discovered great clouds of complex molecules drifting in space: ethylene, formaldehyde, and ethyl alcohol among them.

It's clear the raw materials for life are out there. But what about the right environments? For simplicity, we have to assume it's most likely for complex life to grow and evolve where we did, on planets orbiting stable stars. While there is, as yet, no proof that other planetary systems exist, rings of dust have been discovered circling nearby stars Vega and Beta Pictoris, and many astronomers believe stars like our Sun are naturally born with companions.

Finally, the philosophical tool mentioned earlier, which caps the legitimacy of xenology, is called the cosmological principle, or the "assumption of mediocrity."

Since Copernicus, astronomy has given us a series of lessons in humility, all leading to the conclusion that there is nothing special about where and when we are. First the Earth was displaced from the center of the solar system, then the Sun became a nondescript traveler in orbit about the rim of the Galaxy. Finally, the Galaxy became merely one island universe among billions, and the Universe seems to have no "middle" at all.

The cosmological principle tells us we should avoid the temptation to think that there's anything unique about the Earth in space, time, or situation. Therefore, what has happened here might happen elsewhere, perhaps many times.

## THE DRAKE EQUATION

The most popular way to guess at the possible distribution of technological species was invented by then Cornell Professor Frank Drake when he was at the Arecibo National Radio Observatory. It remains a widely accepted tool for xenological speculation.

Let $N$ equal the current number of technological civilizations in the Galaxy. Then

$$N = R_* f_p n_e f_l f_i f_c L$$

There are reasons to believe that it is short about three factors. But we'll stay with this version for a while. During the 1960s, with plenty of up-and-down leeway in every parameter, Carl Sagan and others estimated that.

$$N = 0.01 \, L$$

This meant the average life span, $L$, of technological races determines the number present in the Galaxy at any time. If self-destruction is the common fate of "civilized" species, there might be no more than a handful in the Milky Way at a given moment. But if a reasonable fraction live long, the Galaxy might teem with intelligent life.

If the planets of a million stars held sophont races, then about one thousandth of a percent of eligible stars in the Galaxy would be orbited by thinking beings. The average distance separating these islands would be several hundred light-years—a gap easily crossed by radio waves.

This was the state of affairs in the early 1970s. The accepted model depicted isolated motes of intelligence separated by sterile tracts of space. Frank Drake and his associates began the search by looking at the nearest candidates. But they found only star noise coming from Epsilon Eridani and Tau Ceti.

Undaunted, they and others expanded the search. Telescopes turned and scanned. The Russians joined in enthusiastically. They too reported only negative results. (In the Soviet Union, extraterrestrial intelligence was not only considered possible, but required by Leninist dogma.)

Astronomers suggested that no advanced species would waste energy broadcasting over the entire radio spectrum. To conserve power and attain a high signal-to-noise ratio, they would modulate over a very narrow frequency band. Yet even the second and third generations of eavesdropping devices, tuned to seek in narrow, so-called water-window or water-hole bands, have come up with nothing so far. Though better instruments are planned, and some radio xenologists are promising vastly improved searches, others have begun glumly proposing that no one is "out there" after all, at least not in our vicinity.

Why this new pessimism? We've only been at the search for less than fifteen years, using spare time and borrowed equipment. According to most calculations using the Drake Equation, the average distance between technological civilizations might be six hundred light-years, or two hundred parsecs. There are well over a million stars in a sphere that wide centered

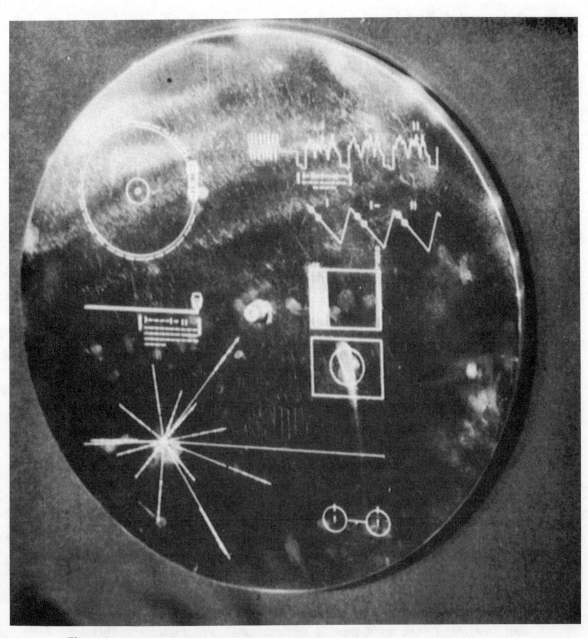

**Figure 1.** The Voyager record, attached to the spacecraft, contains coded information about us, our planet, and our location in the Galaxy. *(Photo: Courtesy NASA.)*

on the Sun. It would take a long time to search even the most likely of these, choosing only those radio bands we guess to be the best.

Two hundred parsecs makes "conversation" a little difficult. But a Sesame Street beacon would be as useful as ever, at that range. And just knowing extraterrestrials exist might profoundly boost poor Homo sap's sagging morale. Success, in the long run, seems assured to the persistent.

What has changed then? What has caused this spreading anxiety?

It's not the sort of thing one would expect to be a cause for pessimism. At first hearing, it sounds like very good news: starships are possible!

## THE THIRD ERA OF XENOLOGY

Sometime in the mid-1970s, several prominent scientists challenged the conventional wisdom that intelligent life arises upon isolated islands, forever separated by the wide gulfs of interstellar space. Ron Bracewell, Robert Forward, and others demonstrated that it's possible, in theory, for spaceships to cross the emptiness between the stars. No "magic" is needed. It isn't necessary to repudiate Einstein. Whether by light sail or anti-matter rocket, humanity may be launching its own starships within a few centuries.

These starships will be nothing like the good old *Enterprise* of television. Limited to possibly a tenth the speed of light, they couldn't travel far by interstellar standards. But they might carry people, possibly several generations, in transit. The "slowboat" generation-ship of science fiction has been mathematically vindicated.

Why is this bad news? Because the possibility of starships presents us with a paradox that is difficult to overcome.

Consider: What would we do if we had starships? If history tells us anything, we would look around for nice real estate and start colonizing. In fact, we wouldn't even need to find planets; stable stars with asteroid belts would do. Professor Gerard O'Neil argues convincingly that cities in space may need little more than raw materials and solar power.

Once the new settlements had reached a high level of industry, they'd send out more colony ships of their own, to stellar systems even further out. Imagine a sphere of human settlement slowly expanding through space. Even limiting ship speed to a tenth of the speed of light, and allowing each colony plenty of time to industrialize, how long would it take for colonies to be planted three hundred light-years from Earth? Ten thousand years? Thirty thousand years?

Mankind has hardly changed in the last thirty thousand years. If we make a few social advances and avoid self-destruction, we should live long enough to fulfill the above scenario.

If *we* could do this, why shouldn't this sort of expansion occur with other sophont species? Isn't it likely many of those million high-tech races we spoke of earlier might also spread and colonize?

If many advanced life forms did this, the 200 light-year "average spacing" between races would be filled up in under 100,000 years!

In fact, all it might take is just one aggressive, colonizing species. Recent calculations by Eric Jones, of Los Alamos Laboratories, indicate a slowly expanding sphere of settled solar systems could fill the entire Galaxy within sixty million years. It's not unreasonable to imagine at least one out of a million civilized races living that long.

Finally, why haven't we met any wise old alien star robots? Frank Tipler, of Tulane University, calculates we should have by now. These robots, first prophesied by the great John Von Neumann, would go exploring like the Voyager and Viking space probes, but would also stop at each solar system to make copies of themselves to send onward. Tipler says just one such self-copying probe could become a horde, with one at every star in the Galaxy in under three million years. Such "Von Neumann machines" should have arrived long ago, and would have been waiting ages for Earth to evolve someone smart enough to talk to.

So we are faced with a new question—why do we see no signs that Earth has been colonized in the last sixty million years? And it's even more curious that we haven't seen signs of civilization near neighboring stars. Why have we picked up no radio signals, when the stars should be humming with information and commerce?

Where is everybody? Does this really mean we're alone?

Debate over this issue has come as a shock to believers in SETI. Just when they seemed to have won acceptance, suddenly they had to contend with jokes that xenobiology was "the only scientific field without a subject matter."

These questions mark the traumatic awakening of xenology as an adult science. It marks the end of a short period of innocence. The dust has not yet settled, but one thing is clear: Some of our assumptions are wrong. The Universe might turn out to be considerably more complicated than the optimistic scientists of the late 1960s first thought.

## THE GREAT SILENCE

We see no evidence for ancient alien cities in Earth's crust. Venus, Mars, and the asteroids appear to be untouched.

Most significantly, Earth, until less than a billion years ago, was populated for eons by only primitive prokaryotic bacteria, teeming in the oceans, with no life on land at all. A visiting starship need not have landed colonists. All they'd have to do is be careless with their garbage, or latrine, and the history of Earth would have been totally different; sophisticated alien parasites would have overwhelmed our primitive ancestors. Since this didn't happen, it seems unlikely aliens landed here during that time. [For a different view on the potential for damage by alien trash, see the articles by Clement; Benford.]

It certainly looks as though we've been alone for ages.

The quandary of the Great Silence has given the infant field of xenology its first traumatic struggle, between those seeking optimistic excuses for the apparent absence of sentient neighbors and those who enthusiastically accept the silence as evidence for humanity's isolation in an open frontier.

Eric Jones and Frank Tipler, in particular, think the apparent absence of ETIS (Extraterrestrial Intelligent Species) simply means this part of the Galaxy is uninhabited. Their "uniqueness hypothesis" implies that some or all of the key factors in the Drake Equation are really very small. For instance, some contend that intelligence such as ours may be an evolutionary fluke.

Then there is Michael Hart's contention that habitable planets like Earth are rare. Alternatively, John Ball dredged up the science-fictional idea that Earth is a "zoo," and extraterrestrials are already here, observing us. (This implies we should add to the Drake Equation a factor to account for ET's *purposely* avoding contact!)

Contact optimists, such as William Newman of Princeton and Carl Sagan of Cornell, have tried to make excuses for the extraterrestrials, suggesting truly advanced cultures would practice zero population growth. The rate of "galaxy-filling" calculated under their conservative assumptions is slow enough to suggest the nearest space-faring race might not have reached us yet.

Or our assumptions for $f_l$ (the fraction of planets on which life originates) might be too high. Although the precursors of life—sugars, amino acids, nucleotides—seem likely to be common, it's possible the next steps might be much harder, requiring some rare catalyst to set off the process.

These and many other ideas have been presented to explain why aliens aren't here. (A complete catalog will be given later in this article.) All the hypotheses offered so far have problems, though. Some seem to contradict the best knowledge we have. Others, like the "zoo" theory, are untestable.

Let's consider one hypothesis xenological speculators have mostly passed up. It's a bit frightening. But maybe not as much as others we'll pick up later on.

## THE FATE OF "NURSERY WORLDS"

In the Drake Equation the combined factor $f_{ic}$—the fraction of life-planets on which intelligence and technology eventually evolve—is generally assigned a value of about one in one hundred. The xenologists who put forward the "one-percent" argument support it by citing the apparent fact that it took four billion years for Earth to give rise to merely one technological race. This is almost half the viable life span of the planet. Intelligent life would seem to be a rare and wonderful thing.

But is this assumption tenable? Let's consider the life cycle of a "nursery world," a planet with a stable biosphere in which the slow evolution to intelligence can take place.

Evolution appears to have proceeded gradually at first, then at an accelerating pace for three billion years. Except for the introduction of sexual reproduction, and later of angiosperms (flowering plants), there is no evidence even to hint the Earth was ever suddenly invaded by extraterrestrial flora and fauna. The Great Silence seems, at first glance, to have stretched through the entire Paleozoic.

If we assume Earth lay untampered with until at least the time of the Jurassic, we can guess that it takes about three billion years for life on a nursery world to evolve to a level of complexity that makes intelligence feasible.

But if humanity suddenly vanished? Would it take another three billion years for intelligence to arise on Earth again? If so, it's reasonable to accept the guess that only one or two technological species could erupt per habitable planet.

But *Homo sapiens* isn't the only species to have benefited from three billion years of evolution. Today's German cockroach may look like his distant ancestors, but he has accumulated many little tricks his cousins in the Triassic never heard of. The size of genome of the raccoon and wolf (the number of genes in their chromosomes) is no smaller than that of man.

Consider what's happened since the Cretaceous-Tertiary Catastrophe, approximately sixty-five million years ago, which wiped out nearly every species of land animal massing more than forty kilos.

The creatures who went on to dominate the planet were small mammals: the early equivalent of mice, and tree shrews. We are among their descendants.

Now, despite the present arms race, man still lacks the ability to exterminate mice. The sudden demise of this star system's current technological race would not finish off the Earth as a nursery. If "mice" did it once they could probably do it again.

Perhaps suitable worlds must pass through long initial "fallow" periods before attaining a level of biological sophistication ripe for intelligence. Afterward, though, such planets might produce sapient species at fairly short intervals, depending on the time needed to recover from the damage done by the previous sentient race.

## EXPANSION SHELLS

It is generally assumed that a space-faring species will expand into the Galaxy because of either raw curiosity or population pressure. For a race limited to slow-boat technology, colonization will take place only in a thin, growing shell surrounding an older, settled region within.

If population pressure is the primary motive for expansion, what of the long-occupied worlds in the interior, especially near the home planet? The words "population pressure" suggest the likely fate of these worlds.

Consider the settlement of Polynesia from roughly 1500 B.C. to about 800 A.D. (The island-hopping analogy is apt.) Eric Jones borrowed growth and emigration rates for his model of interstellar settlement from Polynesian history. The intrepid Polynesians testify to the likely viability of "star-hopping" colonization ventures.

Polynesia may, indeed, be representative of interstellar settlement, but not only in a pleasant sense. The Hollywood image of island life is paradisical, but Polynesian cultures were subject to regular cycles of overpopulation, controlled in war or ritual by culling adult male population. There are stories of islands whose men were wiped out completely.

Meanwhile, introduction of domestic animals disrupted island ecosystems. Many native species were wiped out.

The most severe example is Rapa Nui, also called Easter Island. Isolated thousands of miles from its nearest neighbors, it was as like an interstellar colony as any place in human history. Mankind may devoutly hope to do better when we finally do embark to the stars.

The colonists wrecked the virgin ecosystem of Rapa Nui. When no trees remained to make houses or boats, they had to abandon the sea and it resources, along with all possibility of escape. What remained was native rock—which they carved into hauntingly desolate images—and warfare.

The story of that place should be a lesson to make us all thoughtful.

Now, assume a settled sphere of expansion by an extraterrestrial intelligent species. What of the *inner* systems, within the sphere? The Polynesian example suggests a dismal image of increasing competition for dwindling resources with no escape valve for excess population, since all surrounding systems are in similar straits.

What happens to those inner worlds? In an old, settled system, all available asteroids would long have been turned into habitats. Safe inner orbits with unhindered access to solar power would be at a premium. Even the most efficient space structures will require frequent replenishment of gases such as oxygen, hydrogen, and nitrogen. Comets might supply part of this need, but terrestroid planets would be closer.

One might expect to see a profound cultural split between those living on planetary surfaces and those in space, as depicted by Larry Niven and Jerry Pournelle in their novel *Mote in God's Eye*. It would be simple to bombard cities with redirected asteroids. Factor $L$ (average lifetime) clearly falls in such a case.

In any event, it would be the innocent higher animals who would suffer most in such a crossfire.

## CYCLES OF RECOVERY AND EXPANSION

Recently scientists have uncovered thin layers of clay rich in exotic elements, including iridium (up to five times the normal abundance of some isotopes), at levels associated with the end of the Cretaceous Period. Discoveries in locations from Italy to New Mexico seem to correlate with the great extinction. Some scientists conclude that a major meteorite impact kicked up a great pall of dust, severely altering weather patterns (perhaps in conjunction with major volcanic activity), resulting in mass starvation.

The Cretaceous-Tertiary event wasn't the only one of its kind. At least four other mass extinctions are found in the sedimentary record, including one at the end of the Devonian and another at the Permian-Triassic boundary, approximately 225 million years ago. These events are less well understood and may have taken place over longer periods than the Cretaceous die-back, but we may compare the rough 10- to 500-million-year intervals seen with those suggested by Newman and Sagan for galaxy filling by space-traveling species.

Here's one possibility to consider—might the ecological holocaust of the Cretaceous have been a local manifestation of the death spasm of a prior space-faring race, whose overpopulated sphere of settlement spoiled and self-destructed as the shell of colonization passed outward? It's farfetched of course, but also thought-provoking.

If this were so, all neighboring star systems might also have suffered ecological collapse at the same time. Earth might be the first nursery world to have recovered sufficiently, since the last wave of "civilization" passed this way, to develop a species with intelligence again.

Whether or not the end of the Cretaceous corresponded to the agony of dying star-farers, it may well be that colonizing cultures inevitably leave behind them wastelands empty of intelligence and living voices. If we humans initiate an era of interstellar travel of our own, we may find all around us the blasted remains of such an earlier epoch.

Would we then learn a lesson? Perhaps. But with the ever present opportunities for expansion, those humans who exercise self-restraint and environmental sensitivity toward their adopted worlds will not be able to force this tradition upon those who travel far away to establish newer colonies. A nucleus of selfishness may expand faster than a center of more rational colonization. While some settlers may preserve and protect local ecospheres, cognizant of their long-range potential, others may be rapacious.

Which type will we be? Clearly our environmental record here on Earth is a test. The list of extinct species, some of which might one day have become star-farers, is long and growing longer.

The Great Silence may be the sound of sands drifting up against monuments. It may be quiet testament to the fate of species that allow "population pressure" to be their motivation for the stars.

## THE RETREAT OF THE CONTACT OPTIMISTS

In June of 1984, a new subunit of the International Astronomical Union gathered in Boston, devoted solely to discussing the question of extraterrestrial intelligence. At that meeting the trend of several years continued. The "Contact Optimists" —who have fought hardest to believe we have neighbors in space—continued to beat an organized retreat. They dug in behind fortress hypotheses offering excuses for the tardy, laggard extraterrestrials and explaining their strange failure to appear.

**Figure 2.** Photomosaic of the northern Milky Way. The constellations of Scorpius and Sagittarius are visible at the right. M 31 (the Andromeda Galaxy) is visible in the lower left. This is a close-up view of the left third of the frontispiece sky map. (*Photo: Courtesy the American Museum of Natural History.*)

In so doing, the Contact forces have begun to sound downright gloomy. *Starships are impossible,* some of them declare.

*ETs kill themselves off before they get very far,* others say.

Or *extraterrestrials are pinch-pennies,* who would shrink back from the challenge of the stars.

All this from scientists who once carried the science-fictional banner! Strangely, it is their opponents, the Uniqueness crowd, who now cluster in excited circles at these conferences, chattering about starships and galactic colonization.

Just who are the "conservatives" in an argument like this? We certainly do live in fascinating times.

With the possibility of star travel, and colonization, an average separation of a few hundred light-years starts looking trivial. Not only is the Drake equation no longer complete. We see that it doesn't even predict anything anymore!

When we introduce star travel, the Drake Equation suddenly needs three new factors:

> $V$—the velocity at which an interstellar culture grows into space, pausing to settle likely solar systems and rebuild necessary industry, before again continuing its expansion.
>
> $L_z$—the lifetime of a zone of colonization into which a species has expanded, after which the settled region becomes "fallow" once again.
>
> $A$—an "approach/avoidance" factor, different for each culture, representing a "cross-section for discovery" by contemporary human civilization. (How likely is it we would even notice them? For example, a culture with a preference for settling on comets would never have visited Earth, and might exist undiscovered even now in our solar system.)

By including these factors in the appropriate formula one can try to predict $C$—the probability of contact between human beings and extraterrestrials. The seven factors of the old equation, plus the three new ones, give us a space within which to sort out our ideas, an organizational aid that was missing until now.

All ten probability terms are vital. Contact proponents admit that visits to Earth have been sparse, if they even have happened at all. They merely choose different explanations . . . or excuses . . . for the fact that we have observed no beings from other stars. It turns out the differences of opinion between Contact and Uniqueness forces divide quite simply according to which factor each uses to explain the absence of extraterrestrials.

Uniqueness advocates tend to concentrate on the left and middle of the Drake Equation. For instance, some claim that planets are rare, or that many Earth-like worlds get trapped into Venus-type runaway greenhouse effects, destroying any chance of developing life.

Other Uniqueness savants bitterly dispute this. Planets are plentiful, men such as Michael Hart say, but the odds of independently evolving life are small. Still others, such as Eric Jones of Los Alamos, claim life is probably common enough, but it is the step to *intelligence* that is a fluke here on Earth, unlikely to be repeated elsewhere.

Those on the Contact side disagree, of course. They believe technological societies should crop up all over the place. To account for the apparent absence, Frank Drake and Bernard Oliver hang on to a belief that star travel is impractical. Factor *V* does have its attackers, then, in the face of a tide of popular and inventive starship designs.

And hypotheses abound as to why extraterrestrials might choose to make themselves invisible . . . to avoid contacting us, or to abjure star travel even if it is possible, or to have neglected to settle our solar system in the more than three billion years that it's been prime real estate.

Any of those explanations might work for one or two races . . . maybe a dozen. But if ETIS are as diverse as men and women, the excuses run into trouble. Can the Contact people seriously contend that, out of millions of races, not one would behave as humans do in so many sf novels . . . setting forth in their space Conestogas to settle and alter their new homes?

(True . . . aliens "might think differently than we do." But if there are enough of them, ought not a *few* think like us?)

As always, the most entertaining Contact Optimist is Carl Sagan. Anxious to find an excuse for the missing aliens, and too smart to disdain starships, he has come up with one of the most fascinating explanations. And in so doing he and his colleagues have possibly done mankind a great service.

## NUCLEAR WINTERS

In the Christmas 1983 issue of *Science*, there appeared an article entitled "Nuclear Winter: Global Consequences of Multiple Nuclear Explosions." It has come to be referred to as "TTAPS"—after the initials of its authors, R. P. Turco, O. B. Toon, T. P. Ackerman, J. B. Pollack, and C. Sagan. This historical document has shaken up thought concerning the potential consequences of modern major warfare.

If the models presented by the TTAPS authors are correct even within orders of magnitude, one can only conclude that the arms race between the great powers is a pointless waste of time and money. Even a "limited" nuclear war will devastate the Northern Hemisphere and leave it barren of civilization, nearly devoid of life.

It would not be due to fearsome blasts, nor even lingering radiation. Some fragment of America or Russia might survive those effects. Rather, it would be the dust kicked into the sky by as few as a hundred nuclear ground bursts, or soot from airburst-ignited fires, that would plunge the world into a frigid night from which very little might survive.

At least that's the contention of TTAPS. In the years following, there have been numerous follow-up studies, but no one has successfully disputed the article's orverall conclusions, that nuclear war could severely affect Earth's climate.

Very interesting—and perhaps vital for us all to think about—but what has all this to do with extraterrestrials?

Well, there is reason to believe the nuclear winter scenario had its birth in a struggle to find excuses for absent star-faring aliens!

Consider the position in which Contact proponents like Sagan found themselves. Unable to convince themselves starships are impossible, they had to come up with some universal mechanism to explain how both the number of extraterrestrial species and their rate of expansion could be small enough to explain the Great Silence.

Sagan's answer was to propose the following:

Assume that two types of species achieve technology—peaceful and aggressive. Peaceful races presumably lack the greedy drives that caused humans to seize every opportunity to conquer and spread, here on Earth. These quiet civilizations will expand to neighboring star systems slowly, if at all. So slowly, we can excuse their absence. They just haven't arrived here yet.

Aggressive types would push ever outward, filling the Galaxy as fast as Jones and Tipler contend. But those species must first pass through a dangerous phase—that period between developing nuclear weapons and viable star travel. Sagan says warlike species either cure themselves of their aggressive tendencies, or die.

In other words, the "optimists" are now suggesting the Galaxy is sparsely occupied by long-lived pacifists—who drive their starshipss only on Sunday, presumably—and by the planetary tombs of all the rest . . . species who couldn't learn to control themselves.

But for this rationale to work, there had to be an easily triggered mechanism for destroying civilizations. It must be more powerful than even bomb blasts or radiation ... so compelling that one could envision it happening again and again, to every warlike race that failed to make the transition to a calmer mode of life.

Nuclear winter appears to offer Sagan's brand of Contact aficionado just such a mechanism.

## A MENU OF EXPLANATIONS FOR THE GREAT SILENCE

Why do we seem to be alone?

Each of the explanations offered so far suggests a way to suppress one or more of the factors in an expanded Drake Equation in order to make the overall contact number fit observations of actual extraterrestrials ... so far zero. Let's summarize a list of popular (and not so popular) explanations.

Starting from the left side of the Drake Equation, we begin with some favorite explanations of Uniqueness proponents.

### *Category One:* Solitude

1. Habitable planets may be rarer than astronomers now believe (Suppress factor $n_e$.)
2. Some unexpected "spark" may be needed to initiate life out of prebiotic compounds. (Suppress $f_l$.)
3. The final step to intelligence may require some "software miracle" that makes it far more improbable than currently expected. (Suppress $f_i$.)
4. Insatiable curiosity and manipulativeness, such as contemporary humans display, may be rare among intelligent species. (This effect would obviously suppress factor $f_c$.) As Author Poul Anderson put it: "The puzzle is why we're as bright as we are. Pithecanthropus was doing all right." He proposes that intraspecies selection, especially sexual, became fierce in protohumans, leading to a strange animal that is uniquely clever and capable of fitting itself to live in vacuum or the bottom of the sea.

If any combination of ideas 1–4 are right, we may simply be the first tool users ever to come along. We are the "Elder Race."

*Category Two:* Graduation

5. Technological species may sooner or later discover advanced techniques that make radio and colonization irrelevant. (Still, it is hard to believe any race would abandon the electromagnetic spectrum—radio and light—completely.)

6. Space-faring sophonts might "graduate" to other realms or unimaginable endeavors, coming to look on planets and starships as mere toys. This would set a limit to the period of expansion, though not, perhaps, to exploration.

Either of these scenarios would lower our expected contact cross-section, $A$, with such a civilization. They might also tend to reduce $V$.

*Category Three:* Timidity

7. There might be reasons species develop an aversion to space-flight. For example, Carl Sagan has suggested that the achievement of immortality might make individuals reluctant to take even the slightest risk.

8. As discussed earlier, those species who don't destroy themselves may "cure" themselves of aggressiveness, and so become slow star-farers.

9. Intelligent species might develop a form of telepathy, through mind-computer links, which makes their lives far richer than existence as individuals. If this happened, they might grow reluctant to venture many light-days from the center of their civilization, in order to avoid, in effect, lobotomizing themselves.

Still, it's hard to imagine these notions applying in all cases, which is what we need from a convincing overall explanation for the Great Silence.

*Category Four:* Quarantine

10. Benevolent species may have a tradition of letting nursery worlds lie fallow for long periods, allowing new sentience to be nurtured there.

11. Observers might be awaiting mankind's social maturity, or may have quarantined us as dangerous. A galactic radio club might avoid too early contact, to let us develop our own unique culture, to contribute something new to the galactic melting pot.

12. No listing would be complete without including the far-fetched idea that aliens are already in covert contact with some on Earth. A charming Poul Anderson story depicts Earth's sole "member of the Federation" as an obscure tribe of southwest American Indians.

13. The *low rent* explanation suggests the Earth is simply too unattractive to be settled, or even visited by aliens. For example, Earth life forms rely almost totally on the left-handed isomers of complex organic proteins and amino acids. Other life forms could be right-handed.

14. Finally, it's possible Frank Tipler's imagined self-replicating robots, which should make star exploration cheap and easy for even the timid—might behave just a little differently than Tipler imagined. Perhaps there are hundreds of *friendly* probes, sitting around the solar system, patiently waiting for us. Perhaps we must prove our ability actually to go out there in person before they will deign to say hello.

There is a problem with the quarantine scenarios, unfortunately. All appear to call for cultural uniformity in the Milky Way . . . some way for the pattern to be enforced for billions of years in a galaxy of constantly shifting neighborhoods and star formations. Such a rigid pattern would seem difficult in a relativistic Universe governed by the speed of light.

## *Category Five:* Interstellar Wanderers

Perhaps waves of interstellar wayfarers *have* passed this way. Travel in vast slowboat starships might select for the sorts of beings who *like* living in space, who even come to abandon planet-dwelling as a lifestyle. This could lead to different behaviors.

15. Truly space-borne sophonts might greedily fragment terrestroid planets for building materials and volatiles, having a terrible effect on factor $n_e$, the number of planets that can support life.

16. Alternatively, they might have a tradition of cherishing nursery worlds, protecting them without any desire to use high-gravity real estate.

But we have looked over our asteroid belts in recent years, and they appear to have been untouched since the beginning of the solar system. No one seems to have disturbed them, yet these are the same small bodies such star-farers would covet—which our own grandchildren may be melting and reforming in a century or so.

Looking over our list so far, none of the explanations seems to explain the Great Silence in a convincing way. What's needed is a universal mechanism that acts impartially over long time scales, which would keep the numbers of extraterrestrial species small, or suppress their rates of expansion among the stars.

A few ideas have been proposed that seem to fit these criteria. The reader is warned that some may seem unsettling. If it's any consolation, I'll try to finish with an optimistic scenario ... one that satisfies all the preceding criteria without being nasty.

## *Category Six:* Dangerous Natural Forces

We've already mentioned the possibility Earth might have fallen into a "Venus Trap" ... the runaway greenhouse effect that killed our sister world ... or the perpetual frozen tundra of the "Martian Trap." Here are some other "natural" hazards. Any of them could have disastrous effects on the last four factors of the Drake Equation.

> 17. In its 230-million-year orbit round the Galaxy, our solar system regularly crosses regions of shocked gas clouds and hot young stars. These can be dangerous events. Spiral arms are where interstellar clouds compress to form new stars, and where supergiants end their quick lives in titanic explosions.
>
> Proposed advanced cultures eventually tire of playing galactic roulette and leave the spiral arms for good—setting up in the Milky Way's "halo" of older stars that drift in long, lazy orbits out of harm's way. That could explain why we don't see anybody flying around this part of the Galaxy: Those who *can* leave, do.
>
> 18. Were the dinosaurs really killed off by meteorite or comet impacts, which triggered major changes in the Earth's ecology? If so, were these and other collisions random? It has been suggested that a small dark, companion of the Sun, called Nemesis, or Shiva, orbits far beyond the comet belt, dipping in every twenty-six million years or so to scatter icy and rocky debris into the inner solar system. Alternatively, interactions with the galactic plane, or spiral arms, might trigger such events.

In any case, other solar systems might be in even worse shape than we are, so often smashed by cosmic debris that we're the first to climb up far enough to look around.

19. Our Milky Way may contain objects far more dangerous than mere shock fronts or falling rocks. Radio astronomy shows that many galaxies contain powerful, dangerous jets of relativistic particles, perhaps caused by huge black holes at the galaxies' cores. It's still unclear whether we share this galaxy with a compact version of such terrors, but already there is strong evidence for a black hole, of a few hundred solar masses, near the center of the Milky Way.

## *Category Seven:* Dangerous "Unnatural" Forces

Nature can be malignant, as we have seen. But there are other dangers, as well, dangers that might arise from life itself.

20. Migrational holocausts. This idea was discussed earlier. What happens to planets that are colonized by an expanding interstellar civilization? Unless the settlers leave large parts of their worlds fallow in wilderness preserves, or engage in "uplift" bioenginering of local higher animals, their mere presence is likely to do harm. A world probably cannot serve as a useful nursery of intelligence so long as it's occupied by a space-faring race. When the interstellar tenants finally vacate or die off, it may be a long time before a local species of tool users evolves.

So Earth might be the first nursery world to have recovered sufficiently—since the last wave of "civilization" passed this way.

21. Inevitable self-destruction is another cheery theme mentioned earlier, suggesting that many alien races found themselves where we now stand, on the teetering precipice between self-ruin and self-control; perhaps only a very few make it.

22. From physics and science fiction comes the dreadful notion of "deadly probes," which devastate life among the stars. A particularly paranoid advanced species might not want any potential competition to rise up elsewhere and so might send forth machines like Tipler's self-reproducing probes, but with a nasty edge. Whenever radio traffic indicates that new sentients (like us) have arisen, these robots would home in to destroy the

infection before it spreads. This need only happen once for it to become the status quo, keeping the Galaxy silent and empty for billions of years.

*Category Eight:* A Grasp at Optimism

Is there any friendly explanation for the Great Silence? Isn't there any way the Universe could look the way it does and still let both sides in the debate get their dream—a galaxy with other minds to talk to, and yet still wide open for our great-grandchildren to have adventures in?

I have managed to come up with one.

> 23. The "Water Worlds" scenario. We've spoken of the Venus Trap and of a Mars Trap, which might yank Earth-like worlds toward conditions where life can't exist. This leaves us with the impression that Terra miraculously found itself straddling a narrow fence between two death sentences, and that might be true.
>
> On the other hand, it might not. Recently, Professors Kasting and Pollack have published persuasive arguments that there is a deep valley, a cusp, between the Mars and Venus catastrophes. Within this valley there is another "trap," pulling toward it all planets within its reach. This is the pleasant trap of the Water World.

The existence of life on Earth has had powerful repercussions. It has taken most of the carbon out of the atmosphere and regulated the planet's temperature so that it varies less than the heat output of the Sun itself. One result: the preservation of vast oceans.

If this turned out to be a common phenomenon, let's consider the possibility that the Earth is unusually *dry* for a water world. In other words, what if the vast majority of this kind of planet has far *less* dry land than ours?

Geneticists say that species diversity and rates of evolution depend on the size of the environment involved. It is unlikely that land creatures would develop to the complexity they have on Earth on a world with only island archipelagoes and tiny continents.

That doesn't necessarily mean *intelligence*, per se, is impossible on such planets. After all, dolphins and whales are pretty bright. But it does imply there'd be few places where "hands and fire" beings would develop the technology and basic outlook necessary to take to the stars.

There might be millions of intelligent species out there, ignorant and uncaring about starships, preoccupied with their own oceanic adventures. The result? Envision our descendants setting forth, as Jones and others anticipate. They find no other star-farers, and at first it seems they are all alone. At last, though, they discover other minds ... minds that pose no threat, no danger.

Intelligent whales, or squid, or octopus ... why should they refuse the roving humans' request to make use of local asteroids to build their cities and factories? If the strange-looking bipeds are willing to bring down exciting toys and machines, why not invite them to come take their vacations on the shores of the "useless" little islands, to splash and play and exchange philosophy lazily under the balmy sunshine?

Humans could be the voyageurs—the transporteers—carrying mail and slow philosophical discussion among the water sapients who will only be grateful for the service, of course, never jealous. Our great-to-the-$n$th grandchildren will have their adventures, and serve to tie the Galaxy together.

It sounds like a way to give both sides in our great debate what they want, without having to have a dangerous, malignant Universe, one that's out to get us.

I promised to end on a note of optimism, and I cannot do any better than that.

Now, if only if were true.

# CHAPTER 4

# TUNING IN—
# WHERE TO LOOK
# AND LISTEN

*Assume that extraterrestrial civilizations do exist. Assume further that they communicate across interstellar distances. How do we find them? The Universe is vast: more than a hundred billion stars exist in our own galaxy alone, and there are a billion galaxies or more.*

*Michael J. Klein explains the strategies that radio astronomers use when searching for ETI. Thomas F. Van Horne shows how and why SETI will eventually move off our noisy, crowded Earth and establish observatories in space. Gregory Benford, internationally renowned both as a plasma physicist and a science-fiction author, suggests that we might search our own planet Earth for evidence that extraterrestrials—or their representatives—may have already been here.*

# WHERE AND WHAT CAN WE SEE?

## BY MICHAEL J. KLEIN

### LOOK TO THE STARS

L ooking up at the clear, dark sky, far from city lights, one can begin to appreciate how difficult it is to decide where to look for signals of intelligent extraterrestrial origin. Thousands of stars can be seen with the unaided eye, and thousands of millions can be photographed with large telescopes. Around which ones might there be planets capable of supporting life?

Astronomers estimate that our galaxy contains hundreds of millions of medium-sized stars like our Sun. These stars are sometimes called "good suns" because they are large enough and bright enough to supply the energy, heat, and light required to support life on a planet similar to Earth. They are also smaller than the huge, massive stars that have lifetimes of only a few hundred million years, a time that is probably too short for life to develop to an advanced stage. Good suns shine steadily for billions of years.

Our Milky Way Galaxy is a typical example of one class of galaxies that populate the Universe. The Milky Way is a spiral galaxy; stars numbering in the hundreds of billions form a highly flattened pinwheel that rotates around its center. The disk of stars is about 100,000 light-years in diameter but, except near the center, it is only about 2,000 light-years thick. The Sun is located in one of the spiral arms, about two-thirds of the radial distance from the center. From this vantage point most of the stars appear to lie in a sector of the galactic disk containing the galactic center, which happens to be in the direction of the constellation Sagittarius. In addition to the stars, dark lanes of light-absorbing interstellar dust are distributed along the galactic plane, which limits our view to a few thousand light-years. These clouds of gas and dust are known to be birthplaces of new stars and, most likely, new planetary systems.

Many SETI scientists believe that extra time and effort should be spent searching for signals along the plane of the Galaxy. They correctly argue that stars, including many good suns, are concentrated there. An alternative point of view contends that stars in the solar neighborhood are uniformly distributed out to a distance of a few thousand light-years. Within that sphere we find several million stars that appear to be evenly scattered across the sky. Putative signals from planets around these stars will tend to be stronger because they are closer to us. The numbers per unit area of sky favor the search along the galactic plane; the potential signal levels favor the solar neighborhood, which also includes a lot of good suns. The best bet is to try to accommodate both approaches.

The galactic center is so densely populated with stars that one might be tempted to concentrate a search in that single direction. However, the galactic center does not appear to be a very friendly place for life, at least not for life as we know it. Intense bursts of gamma rays, x-rays, and energetic particles have been detected from the galactic center. A growing number of astronomers now think that a massive black hole, millions of times more massive than our Sun, may be responsible for the exotic radiation that is observed.

## THE COMMUNICATION MEDIUM

Even if we knew the precise location of an ETI civilization, how can we know what method of communication they might choose to use? What form of energy will effectively carry information across the vast distances between the stars?

Nearly everyone agrees that photons are the answer. Photons are the massless packets of energy that characterize the particle nature of electromagnetic waves, which we recognize most easily as visible light. Other electromagnetic waves include gamma rays, x-rays, ultraviolet and infrared light, and radio waves. The only difference among these various electromagnetic waves is their frequency, which is related to the energy of the photons. They all propagate through the vacuum of space, and most are easy to transmit and to receive. Even the simplest forms of matter, electrons for example, require at least a thousand million times more energy than photons to achieve speeds approaching the speed of light. Photon transmission is thus the speediest and most effective means of interstellar communication we know about. If we are correct in this analysis, for which frequencies (or energies) should we design and build our telescopes and detection equipment? Might there be *preferred* frequencies for interstellar communication?

Visible light is a candidate with certain advantages for interstellar communication. Our Sun, just like all the other stars, shines at all frequencies across the electromagentic spectrum. The peak of its brightness occurs in the visible part of the spectrum. Depending on its size and age, a star's brightness peak will be shifted a bit toward either the ultraviolet or the infrared, but the bulk of the energy from the class of good suns will be in visible light. It is possible that life evolving on planets orbiting these other stars would develop some form of "eyesight" that would match the frequencies of their stars' brightness peaks, just as we have. In other words, we all might have visual sensors at similar, but not identical, wavelengths. Might this characteristic be recognized as something common to all sentient beings and exploited to establish communication?

Instruments to search the sky at visible wavelengths are available, but to date little has been done to use the world's great telescopes expressly for SETI. This may be due to the fact that optical wavelengths do not appear to be the best choice from an instrumental point of view. The problem is competition! Anyone who designs (or pays for) a transmitter for interstellar communication would surely realize that a signal transmitted with a given power at almost any wavelength *other* than visible light would be more detectable at the receiving end than the optical signal, which could be lost in the glare from the star. For example, high-power microwave radar transmitters can literally outshine the Sun a billion times, while a state-of-the-art optical laser system, with similar effective radiated power characteristics, would just match the Sun's brightness or at most exceed it by a factor of ten.

Figure 1 shows the solar brightness curve compared with realistic signal powers that we could transmit at three frequencies (one microwave, one visible laser and one infrared laser frequency).

SEARCH STRATEGIES: WIDE VERSUS NARROW FIELDS OF VIEW

For the moment, let us assume that we can correctly guess the frequency of the signals we are seeking. A systematic search near every good sun in the Galaxy, star by star, could easily last several lifetimes. How can the search be accomplished in a reasonable time span of a few years? The answer depends on the kind of telescope selected for the task. As we gaze up at the sky, we can see a good fraction of a hemisphere at one time. When we use a telescope, our view of the sky is usually limited to a small patch of the sky. The telescope is billions of times more sensitive than the eye, but the field of view is highly limited. In general, the greater the sensitivity, the more restricted the field of view, and this holds for the entire electromagnetic spectrum, from the shortest gamma rays to the longest radio wavelengths.

Modern SETI projects require telescopes and signal processing equipment specifically designed to detect a signal, perhaps a very weak signal, in the presence of a flood of background noise. Of course the signal might *not* be weak. A highly advanced technological civilization might be inclined to transmit signals so intense that they would be obvious to anyone who happened to tune in. The fact that we have not yet discovered such signals perhaps suggest that the task will not be that easy, however.

Interstellar distances are so enormous that even the most powerful transmissions die away to a whisper as they propagate across the Galaxy. Earth's most powerful radio transmissions, which are routinely received by our spacecraft at the edge of the solar system, are fifty million times weaker when they finally reach the distance of the closest stars some four years later. These same transmissions would be a million billion times weaker if they were to be received at the center of the Galaxy. In SETI we are searching for signals; we are not transmitting. We do not know how far away the nearest transmitting civilization might be. To increase our chances for success, SETI instruments should be as sensitive as our technology (and budget) will allow.

## GUESSING THE SIGNAL CHARACTERISTICS

Anyone who has ever been unexpectedly separated from a friend in an overcrowded amusement park has, in a small way, experienced the predicament of SETI scientists. Each person tries to guess what the other might do to solve the problem of finding each other in the crowd. The chances are much improved if the friends know each other very well. The size of the park also makes a difference. The challenge for SETI is to guess what "they" might do to make contact without benefit of knowing anything about them, without knowledge of where they might be located in the vastness of galactic space, and without information about the kind of signal they might choose to send.

Nevertheless, we are forced to make decisions if we wish to conduct an active search for evidence of extraterrestrial technology. Speculating among ourselves about what might or might not be true costs little; building instruments and equipment to carry out a scientific experiment can be expensive. Following the example of space research projects, SETI scientists and engineers must consider a variety of potential approaches and finally make choices, often difficult choices, to develop an experiment that will have the best chance of success and yet be affordable to build and operate.

To date, the vast majority of search space that might contain ETI signals remains largely unexplored. The reason is that there is so much to explore! In this context, search space means something beyond the three-dimensional volume of the Galaxy. It is a multidimensional space that includes source location, signal frequency, power level, time of arrival, signal modulation, and polarization. ("Signal modulation" and "polarization" are technical terms that respectively describe how a signal changes with time to carry information and how it vibrates in a plane perpendicular to the direction the wave travels.)

Plausible limits for these "search space" parameters may be impossible to define. It is clearly not practical to look for all possible signals at all frequencies from all directions using instruments with the greatest sensitivity that our technology can provide. Each reduction of search space not only decreases the search time and cost, but also decreases the chances for detection. Therefore, tradeoffs must be carefully made between sensitivity and signal character on the one hand, and spatial directions and frequency ranges covered by the search on the other. The objective is to select the most effective strategy within the constraints imposed by the limits on time, resources, and technology.

First we assume that the signals, if they exist, will be intentionally transmitted for detection by other technological civilizations. In other words, we assume that the signals are of a type that we are capable of detecting. If they do not wish to allow others to tune in, then it is unlikely that we will intercept their messages. We might look for "leakage" signals, an alien's version of our own commercial broadcasts or our military radar transmissions, which are expanding at the speed of light away from Earth within a growing sphere (now about ninety light-years in radius). Leaked signals could be very hard to detect because they would probably be much weaker than beacons, and they might be unrecognizable to us. This is especially true if the senders are even slightly more technologically advanced than we are. Our eighteenth-century ancestors never could have detected twentieth-century television signals had they been sent to them with the most powerful transmitters of our day.

We assume that we are the technological infants who, for the near future anyway, will be searching by listening and looking. The transmitting societies must at least equal our technological level, for only in the last fifty years have we ourselves developed the capability of transmitting across interstellar space. If their technology happens to be far in our future, then we must rely on their desire to announce their presence to newcomers like us.

If a transmitting ET society wants its beacons to be detected, then what kind of signal would they choose? A logical approach would be to transmit signals that will stand out against the natural background emissions from the Galaxy and beyond. We look at the night sky and see the moon, the stars, and our neighboring planets. But our astronomical instruments see much more. They pick up intense emissions from galaxies, quasars, huge interstellar clouds of gas and dust, black holes, and even the faint afterglow of the cosmic Big Bang that began it all. These sources of cosmic emissions are detected at all frequencies across the electromagenetic spectrum. However, the intensity is not the same at all frequencies.

As noted above, visible light would surely catch our attention if we could separate an artificial signal from the overwhelming glare of the starlight. This problem of contrast is very difficult to overcome. Furthermore, signals at the highest frequencies, which include the gamma rays and ultraviolet as well as visible light and the near infrared, are absorbed and scattered by interstellar gas and dust clouds. This feature makes them less effective for interstellar communication than the far infrared wavelengths or the radio frequencies. Radio frequencies not only travel unhindered from the farthest parts of the Galaxy, they also happen to fall in the part of the spectrum where the background emissions from natural sources have minimum intensity.

The subject has been extensively studied and debated in the scientific literature and at professional meetings involving scientists from many disciplines. A systematic search of the radio frequency portion of the spectrum has emerged as the strategy of choice for the detection of extraterrestrial signals.

## THE MICROWAVE WINDOW

Even with a decision to begin the search with radio frequencies, the number of possible frequencies is still too large to cover economically. However, there are natural boundaries to the optimum radio frequencies for interstellar communications. There exists a natural "microwave window" that would be known to all technological civilizations in the Galaxy regardless of where they happen to reside. As the name implies, the window occurs at the so-called microwave frequencies, which coincidentally are used here on Earth for communications; we even use them for terrestrial culinary purposes. (See Figure 2.)

On a graph, with background noise plotted on the vertical scale and frequency plotted along the horizontal scale, the microwave window appears as a broad valley of low background emission lying between two steeply sloping "cliffs." The lower frequency wall is the result of intense emission from energetic electrons spiraling in the magnetic field of the Galaxy. This wall of radio noise rises steeply at frequencies below about 1,000 megahertz (MHz) or one gigahertz (GHz). The latter happen to be the frequencies we use for television, which work just fine here on Earth but would have to compete with the galactic emissions if we tried to transmit across interstellar distances.

The upper boundary of the window is caused by the so-called quantum noise that plagues all high-frequency amplifiers. This spurious noise grows stronger with increasing frequency, but the impact of this effect occurs well above the frequencies that are completely blocked in our atmosphere by oxygen. This means that the terrestrial microwave window, accessible from the ground, is not as wide as the free-space window that can only be utilized above Earth's atmosphere. Earth's atmosphere also contains water vapor, which further restricts the reception of high frequency microwaves. In the final analysis, the practical limits to listening for an ET signal in the terrestrial microwave window lie between 1 and 10 GHz. (Inside the window, the faint background noise is due to the universal blackbody radiation, the remnant of the Big Bang.)

## THE WATER HOLE

The microwave window happens to include a special set of frequencies that has been identified as possible "signposts" for interstellar communication. Most of the background radio noise from the Galaxy is uniformly spread over wide ranges of frequency, rather like the static we hear from a radio tuned between transmitting stations. Several decades ago, however, excess radio noise peaks were observed by astronomers at a few discreet radio frequencies. The source of the noise was identified as spontaneous emission from atoms and molecules located in cool clouds of gas and dust distributed in the spiral arms of our Milky Way Galaxy. Each atom or molecule, upon excitation, randomly emits a short burst of energy at a very specific frequency. The random emissions from billions of atoms and molecules are detected as excess noise at a single frequency, which is different for each type of atom or molecule. Plotted on a graph, the excess emission appears as a spike rising above the background noise continuum. (Atoms in hot clouds showered by stellar ultraviolet emit in a similar way, but they exhibit bright emission lines at much shorter wavelengths; for example, in the visible or ultraviolet part of the spectrum.)

The process is a bit like crickets chirping in a large field. A single cricket might not be heard, but thousands of these randomly chirping insects can raise quite a din, and if they could all chirp identically at the same frequency, the sound would be even more noticeable.

To date several dozen molecules have been identified as sources of interstellar radio line emissions. Among the first radio lines detected were those due to neutral hydrogen atoms (at 1,420 million cycles per second or 1.42 GHz), the most abundant atom in the Universe, and the hydroxyl radical ($OH^-$, with several lines near 1.65 GHz). These two species are widely observed by astronomers to map the spatial distribution and motion of matter in the Galaxy. The frequencies of these particular lines, coincidentally, happen to be close to each other in the quietest part of the microwave window.

Astronomers anywhere in the Galaxy would surely know about and study these line features just as we have. They might also notice that the two fragments are the constituents of the water molecule and that they are liberated when broken apart by intense ultraviolet radiation or by collisions with other molecules. If water is a uniquely critical component for life, might these facts about the radio lines of these molecular fragments lead others to use them as signposts for interstellar communication?

No agreement about the answer to this question has been reached among the SETI scientists here on Earth. The subject of preferred frequencies has been discussed and debated for years without resolution. A wide variety of special "magic" frequencies have been suggested, but none has captured the imagination of the majority. Nevertheless, the "water hole" frequencies discussed above tend to be included in most SETI plans.

# COMMUNICATION SIGNALS COMPETE WITH THE SUN

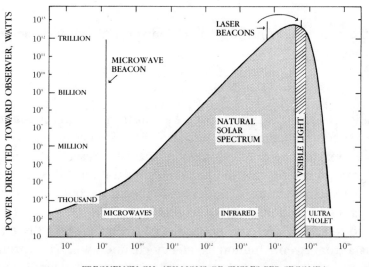

**Figure 1.** If a microwave, visible laser, or infrared laser beacon were directed at an observer from near a Sun-like star, it would appear as a spike against the natural spectrum as shown above. *(Diagram designed by M. Klein and W.R. Alschuler. Art by Elizabeth Wen.)*

## THE MICROWAVE WINDOW
## FROM EARTH'S SURFACE & FROM SPACE

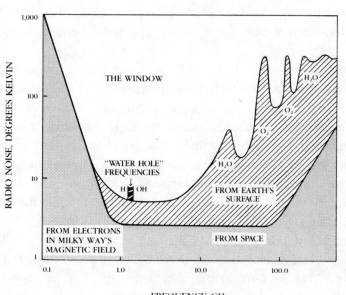

**Figure 2.** The dotted area represents the natural background of microwave noise in space. The hatched area represents noise contributed by Earth's atmosphere. The magic frequencies in the "water hole" are represented by the stripes. *(Diagram: Michael Klein, W.R. Alschuler. Art by Elizabeth Wen.)*

Perhaps the reasoning is based as much on romantic hope as upon scientific rationale.

## THE NASA SETI PROJECT

The National Aeronautics and Space Administration is preparing to carry out the most comprehensive search to date. The project is a ten-year effort that will use existing radio telescopes and specially designed receivers and digital signal processors to conduct a highly automated search of the microwave radio spectrum. Using state-of-the-art technology, wide-band receiving systems and high-speed digital processing systems are being constructed to search the "cosmic haystack" far more efficiently than ever before. Data from twenty million frequency channels will flash through the digital processors as custom-designed microprocessor "chips" and commercially available electronics process the data at blinding speed. Several dozen of the most powerful general purpose supercomputers (e.g., the CRAY X-MP/18) would be required to match the speed of this special purpose hardware, which will complete tens of *billions* of mathematical computations each second.

The automated receiving system will simultaneously search for signals in twenty million frequency channels and record anything unusual that stands out against the background noise. Follow-up observations will then be made to determine which, if any, of the reported "events" are potential ETI signals.

The NASA Microwave Observing Project will follow two complementary search strategies: a high-sensitivity examination of nearby solar-type stars, designated the "targeted search," and a complete "sky survey." The two approaches are complementary in that the targeted search stresses sensitivity to detect very faint signals that could be either pulsed or continuous, and the sky survey gives up sensitivity in order to survey the 99 percent of the sky that is not covered by the targeted search.

The primary task of the targeted search will be to examine nearby stars that are similar to our Sun. Approximately eight hundred of these stars have been identified within a radius of 80 light-years, which is the limit of the current star catalogues. Automated receiving systems designed to search for signals at 1 to 3 GHz in the microwave band of frequencies (which includes the water hole) will be used with the world's largest and most sensitive radio telescopes. The 305-meter-diameter antenna at Arecibo will produce the most sensitive search, covering the one-third of the sky that

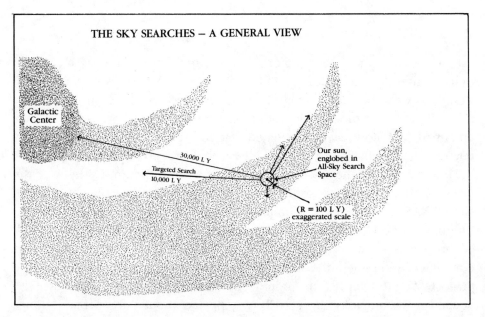

**Figure 3-A.** The Sun, in a spiral arm of the Milky Way, englobed in the all-sky search space. *(Diagram: W.R. Alschuler. Art by Elizabeth Wen.)*

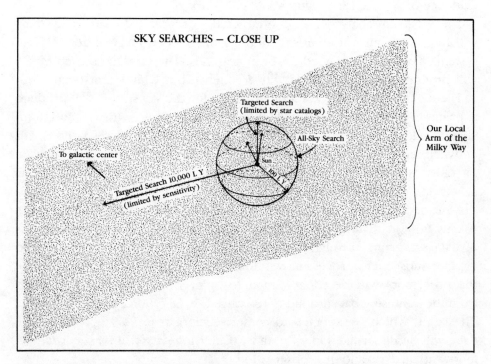

**Figure 3-B.** Close-up view of the SETI search space, and the limitations of each approach. The all-sky search is limited by the sensitivity of its receivers. The targeted search is limited to catalogued stars. *(Diagram designed by W. R. Alschuler. Art by Elizabeth Wen.)*

can be observed at that equatorial site. Other large antennas located at radio astronomy observatories and at NASA's Deep Space Communication Complexes will be used to complete the targeted search at more northerly and southerly latitudes. The radio telescope at each site will be programmed to follow each designated star for several minutes at each frequency, while the targeted search analyzer looks for complex signal patterns.

The objective of the sky survey will be to search the entire sky over the primary frequency range from 1 to 10 GHz. The 34-meter-diameter antennas of NASA's Deep Space Communication complexes will be used with support from other antennas of comparable size and sensitivity. These will be programmed to scan the sky in predetermined patterns calculated to complete the survey in the six years allotted to the operational phase of the project. About twenty-four months of scheduled antenna time will be needed. The entire sky survey will be repeated about thirty separate times as the receiving systems are stepped through a series of adjacent frequency bands, each 300 MHz wide. As each scan progresses, power levels in each of the twenty million frequency channels will be automatically analyzed for evidence of signals. (See Figures 3-A and 3-B.)

The spectrum analyzers for the targeted search and sky survey systems will be similar but not identical. MultiChannel Spectrum Analyzer (MCSA) systems have been designed for both searches. The targeted search MCSA will produce multiple outputs with frequency bin widths that range from ½ Hz to a few tens of Hz. A modified MCSA for the sky survey will produce a primary output bin width of about 30 Hz, along with a continuum channel (for comparison) about 1,000 Hz wide. These resolutions are between ten and a thousand times finer than the resolutions currently used for radio astronomical research (which looks at gas and dust clouds in our galaxy and external galaxies).

A major feature of the two MCSAs is the requirement that they process twenty million channels of data "on the fly." With twenty-five billion bytes of data screaming through the processors each second, trying to record data for subsequent processing would be totally impractical. Consequently, the processors will be programmed to differentiate potential ET signals from the confusing background of terrestrial signals and cosmic radio noise. The latter, which is expected to comprise more than 99.9 percent of the data, will be discarded. Only the most promising data will be recorded for follow-up analysis and/or reobservation.

The SETI processors will also be used for traditional radio astronomy purposes. Data with potential value for astronomical research will be saved

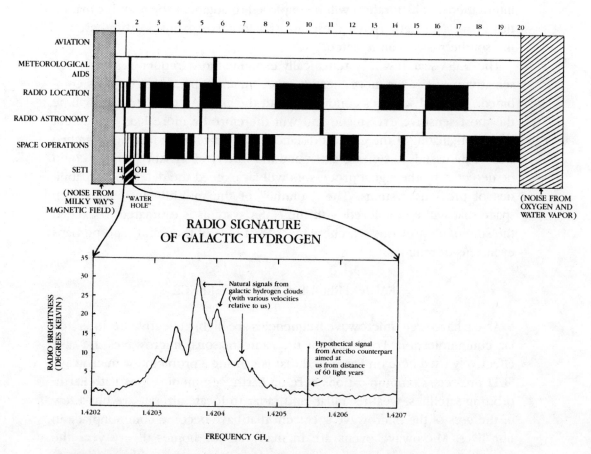

**Figure 4.** The frequency allotments to terrestrial and space activities, and the "water hole" frequencies are pictured at top. Below is an expanded view of typical natural radio noise observed from interstellar hydrogen. Superimposed there is a hypothetical ETI signal, represented as a spike. *(Diagram designed by W. R. Alschuler. Art by Elizabeth Wen.)*

during the course of the SETI observations. Plans are being developed to make the processors available for radio astronomers to use on a non-interference basis.

Radio telescopes in the Southern as well as the Northern Hemisphere will be scheduled for the SETI observations. Approximately one-third of the sky survey area, which includes a similar fraction of targeted-search stars, cannot be seen from northern observatory latitudes. Opportunities for international collaboration will be explored to augment the search capabilities in the Northern Hemisphere. Australian participation in the search of the southern sky is anticipated.

The NASA search will systematically examine more frequencies with high sensitivity and cover more "search space" than all previous searches combined. Targeted searches conducted with the giant Arecibo antenna will be the most sensitive ever made and will therefore be more likely to detect signals originating at the greatest distances. The range of frequencies searched will be thousands of times greater, and the variety of signal types that could be detected by the signal processors will far exceed the detection capabilities of previous systems. The "volume" of the multidimensional search space that will be explored with the NASA systems is estimated to exceed the sum of all previous searchers by at least tens of thousands and perhaps even tens of millions.

### RADIO FREQUENCY INTERFERENCE

As we have seen, microwave frequencies are highly effective for interstellar communication. For many of the same reasons, microwaves are also effectively used here on Earth, and that is causing a problem for microwave SETI projects. Communications here on Earth, communications with Earth-orbiting satellites, weather radar, and radar to locate aircraft are but a few of the uses of the microwave spectrum that have become commonplace in our lives. Microwave ovens are in many of our homes. Every year the microwave spectrum becomes more and more crowded with a variety of signals (See Figure 4.) The extremely sensitive receiving systems required for SETI are no longer able to filter out the cacophony of radio noise that permeates the sky. Some of these signals are so intense that they effectively block out weak signals over a wide range of frequencies. From the SETI perspective, these signals are designated as Radio Frequency Interference (RFI).

It would be convenient if one were able to turn off these other transmitters while SETI observations are in progress, but that approach works only in a very few cases. The transmitters are licensed to operate at their assigned frequencies. It is not their fault that their transmissions interfere with SETI. So it is up to us to design our SETI instruments to avoid as much of the RFI as possible. Frequencies where RFI is especially intense may be lost for SETI here on Earth. Special filtering techniques have been designed to deal with this problem.

Studies of the impact of RFI on STEI indicate that the interference levels are serious and that the problem is growing worse each year. SETI researchers are faced with the problem of trying to hear a pin drop in a large ballroom filled with chattering guests, and more guests are arriving all the time!

## SPACE-BASED SETI

To keep the cost at a modest level, the NASA SETI project will be carried out from Earth's surface. However, SETI from space has some very attractive advantages. Free from absorption by water and oxygen in the Earth's atmosphere, space-based SETI telescopes would be able to search the so-called free-space microwave window that extends far above the highest frequency of the terrestrial microwave window. The free-space window extends from 1 GHz to 100 GHz. Frequencies above 10 GHz, which are partially blocked by the atmosphere, should be explored in the future. Depending upon what is learned from Earth-based SETI, it may be important to search these frequencies from space at some future date.

Space-based SETI would also open up new possibilities at frequencies far higher than microwaves. Several SETI scientists have suggested that searches at infrared wavelengths might be effective. These frequencies are difficult to search from the ground, and the receiving systems are currently less sensitive than the microwave systems. Advancing technology may very well bring space-based infrared SETI into the realm of affordable SETI projects of the future, however.

Other researchers have noted that the moon is also an ideal base for many kinds of astronomial research, including SETI. If manned lunar bases become reality, then one can imagine that astronomical observatories could be established on the lunar far side, which would be perpetually shielded from interference from Earth. (Orbiting observations would, unfortunately, still be subject to terrestrial RFI.) International agreements would have to be signed to protect the far side from electromagnetic contamination due to, for example, satellite or lunar-base communications. But if protection could be achieved, large areas of the lunar far side could be maintained as research "parks" free from interfering signals across the entire electromagnetic spectrum. This would allow new classes of scientific research to develop.

Given the wide range of research possibilities, we ought to examine the feasibility of far-side observatories in detail and, if appropriate, build them.

# THE FUTURE OF SETI

## BY TOM VAN HORNE

N
o matter what types of radio search systems may be used in the
next few years, one factor is likely to dominate SETI's future.
Radio Frequency Interference (RFI) is increasing around the
world at an enormous rate, and any SETI project will either have to escape
from it or learn to live with it. What with microwave data transmissions,
cellular phones, police radars, aircraft transponders, and communications
satellites, we are awash in a sea of intelligently generated signals. For SETI,
this means we aren't just looking for a needle in a haystack. Much worse,
we're looking for a slightly odd needle in a haystack full of them.

SETI listening programs are based on the techniques and instruments of
radio astronomy. Yet even while the technology to receive and recognize
ETI signals is rapidly advancing, the increasing volume of human-made
noise is slowly crippling the world's great radio telescopes. There is serious
concern that conventional Earth-based radio astronomy may only have
another decade or two of existence before satellite transmitters, the most
severe offenders, have obscured the entire radio sky.

Steps to preserve access to the sky and to restrict the extent of interference, at least at certain frequencies, are being taken by several groups
concerned about astronomical research. But in the face of powerful commercial concerns, weak international enforcement powers, and military
planners who think restricted frequencies are a good way to hide secret
transmissions, there seems little chance that anything will stop the growth
of RFI on and around Earth. To continue to operate at all, future SETI
systems will either have to move beyond the range of human interference
or else somehow learn to see through it.

The only way to avoid terrestrial RFI is to set up operations beyond Earth, which for the foreseeable future will be much more expensive than operating on Earth. Still, space-based SETI offers a number of advantages to counter higher supporting costs. Freed from the strain of fighting Earth's gravity, orbiting SETI radio telescopes can be built with concentrating reflectors that are much larger than is possible on Earth. A large spherical reflector in orbit would only have to be strong enough to hold its shape against the weak "tidal" stresses felt in orbit. It could be structurally very simple and could focus into a large number of free-flying feed horns and receiver systems to study a number of targets simultaneously with great sensitivity.

A SETI facility beyond near-Earth orbit would be able to protect itself from terrestrial RFI by means of a simple screen held in position between the reflectors and the Earth. The schedule of observations and choice of targets for such a facility would also be much more flexible than for a ground-based dish. Separated from the obscuring bulk of the Earth itself, the field of view of an orbiting antenna system would include the entire sky at all times, except for the small area obscured by its protective RFI screen. As mankind engages in more and more space-based activities, it becomes hard to imagine that spaced-based radio telescope facilities won't be constructed. It seems likely that such telescopes will ultimately become home to SETI research projects. The only question is when.

A variation on free-space antenna systems that also escapes the Earth's RFI environment is to use an enormous natural radio screen—the moon. Near-Earth orbiting radio telescopes are still vulnerable to satellite RFI from sources in higher orbits than themselves, especially Earth's noisy ring of geosynchronous communications satellites. But because Earth's moon is tidally locked to our planet, the far side is protected from the entire area of terrestrial radio interference by a thousand miles of solid rock. There has been considerable discussion of using the far side of the moon as a haven for astronomy, associated with proposals that the U.S. and other nations establish a permanent manned presence on the moon. If manned operations are being supported on the moon for other purposes, it may well be that the then existing lunar program and the natural shielding might make radio astronomy from the far side attractive. Large-scale reflectors could be constructed with a design similar to the Arecibo reflector, using some of the many impact craters on the lunar surface. The rims of the craters would be used to suspend cable supports that would give the large dish reflectors their shape. The lower lunar gravity would make possible structures very large by terrestrial standards.

There are also a number of disadvantages to far-side radio astronomy. Reflective dishes in lunar craters can't match free-flying dishes either for size or for flexibility of operations. Most important, it is unlikely that the entire far side of the moon will be devoted solely to astronomy. If any other human activities, even exploration, are being carried on in that hemisphere, some form of communications link will be necessary. Most probably, communications satellites would be used for those links, and at that point the shielded environment of the lunar far side wouldn't be so shielded anymore.

An alternative to escaping from the noise of human civilization is to learn to live with it. For years, radio observatories have used ingenious technical tricks to limit RFI problems. Digital signal-processing technology offers some hope that advanced radio telescopes may be able to "subtract out" signals, if they can be identified as RFI.

The future of SETI is tied to what our technology can do. Over time, as our technology advances, new signal-processing techniques and new telescope designs will become available. Entirely new systems will revolutionize our search efforts. One such breakthrough system may be emerging from work currently underway by the Ohio State University SETI project.

OSU SETI has been developing a new technology known as the "radio camera" as an astronomical instrument. A radio camera uses a phased array of antenna elements, receivers, and digital processors to create an image of the radio sources in its field of view. In a standard phased-array antenna system, like those used by large strategic defense radars, the signals from a number of small antennas are combined to "build" an antenna beam like that of a single large antenna. The direction of that beam is determined by the way in which the signals are processed rather than by the physical orientation of the antennas themselves. Thus, a phased-array missile radar can "point" electronically to scan the sky much faster than a dish antenna could be mechanically moved to cover the same area. A radio camera takes advantage of the fact that a digital phased array is not only able to "build" a beam pointing in any direction but also can use the same set of data to "build" every possible beam and "point" in all possible directions at once.

The OSU SETI project to develop an astronomical radio camera is named Argus. The goal of Project Argus is to create a radio telescope that is capable of imaging the entire sky *simultaneously*. Project Argus would use a phased array, probably consisting of small helical antennas about one foot in diameter by one foot tall, laid out in a spiral pattern perhaps a few hundred feet in diameter. Each antenna would be connected to its own receiver and digital-processing package. These would send the digital data from each element to a central computer for storage and processing. The computer would then use that information to "build" telescope views of the entire sky mathematically.

The performance of an Argus system is determined by a number of factors, and there are many trade-offs involved. Its angular resolution could be much better than a dish antenna of the same diameter as the spiral. Sensitivity will be determined by factors such as the total collecting area of the array, receiver sensitivity, bandwidth being observed, and time of observation devoted to each sky direction. A thousand-element Argus would still have a small collecting area compared with a large dish like Arecibo, but because it would look at the entire sky rather than tiny pieces of it, Argus could use very large integration periods—the radio equivalent of long film exposures—to make major gains in sensitivity. The greatest trade-off for an Argus system is likely to be in bandwidth. The limiting factor in Argus is the sheer volume of digital data to be produced, stored, and processed, and bandwidth is the greatest single factor in determining the volume of that data. Argus, at least initially, would have to create a series of views of the sky using different frequency bands. However, as information processing technology advances and our ability to store digital data increases, this limitation will be greatly diminished.

The end result of all this would be rather like having thousands of large radio telescopes looking in every possible direction all the time. This can eliminate the entire question of "where to look"—Argus would look everywhere. An Argus SETI system would fill one entire slice of the "cosmic haystack" just by being turned on.

OSU's Argus and similar "radio camera" systems may also be able to "see" around RFI entirely. Radio sources near Earth, be they ground-based, airplanes, or satellites, because they are tied to the Earth, appear to move differently than do sources at astronomical distances. Argus would form an image of where every radio source appears in the sky, and over time this would be a record of how those sources seem to move. Since Argus looks at everything, it doesn't really care whether something is interference or not. So long as its receivers are able to handle the total amount of radio input from the sky without reaching saturation, Argus could only be blocked along the line of sight of the source of interference. The sky would have to get more crowded than we can imagine to disable an Argus telescope.

**Figure 1.** Orbiting radio telescope with RFI shield, as proposed by NASA. *(Art: Courtesy NASA.)*

**Figure 2.** Using the body of the moon as a shield from terrestrial interference, a crater dish on the lunar farside could be the best location for radio astronomy. *(Art: Courtesy NASA.)*

Eventually, advanced multidirectional systems like Argus will be set up in space. The digital data they will store as they observe the Universe will be examined for intelligent signals just as it will be examined for information about the distribution of matter in distant galaxies. If we still have found no evidence of intelligence other than our own, we will have to make decisions. Will we begin broadcasting the messages we hoped to hear? Will we begin the development of the self-replicating intelligent probes that we failed to find (or that failed to find us)? Any discussion of the future of SETI must be pessimistic, because if the search is successful, everything about the field will change in ways we cannot hope to predict. But it may well be that before the systems being planned today are off the drawing board, we will have our answer.

# ALIEN TECHNOLOGY

## BY GREGORY BENFORD

Interstellar travel is a prodigiously expensive proposition. Sending a manned expedition to a nearby star would take about a thousand times the total energy now used annually in the United States. Though colonization of the planets seems a plausible long-term goal for us, an interstellar expedition would take about a hundred million times more energy than establishing settlements on Mars.

Numbers like these lead many to assume that no alien technological society has *ever* carried out far-flung exploration of the Galaxy. Though surely any older race would have to expend vast reserves to send ships to our star, we should be careful in ruling out the prospect. We humans built the pyramids and climbed Mount Everest with no rational motive and at high cost. Of course, we have also abandoned great projects, like manned missions to our moon. Similarly, for decades we have had the technical ability to build a truly spectacular building a mile high, but we don't. Judging the limits of the grandiosity of a species, even our own, is not easy.

Also, there are dangers in such assumptions, because they can stop us from considering how first contact might conceivably occur locally, near Earth. This idea animates the greatest of science-fiction films, *2001: A Space Odyssey.*

True, we have found no monoliths or other records of alien visits—at least, none we can seriously credit. But this may come from our lack of imagination, our inability to look in the right places. I think it is worth the time to seriously, dispassionately look into the possibility that intelligent beings may have visited Earth, or ventured into the solar system at some time in the distant past.

Thinking about this question in a systematic way is not easy. We are trained to believe that good, hard, no-nonsense thinking is the best way to attack problems. But here something different is needed—soft thinking, if you will. By this I mean the ability to speculate but remain within boundaries.

After all, any extraterrestrials who visited Earth possessed technology (and perhaps wisdom) far beyond our own. We should be properly humble about what such beings could do. This demands mental flexibility, to say the least.

It might well be that a race capable of journeying among the stars will possess a technology we would not even *recognize*, much less understand. Add to this the fact that these beings are truly alien, in all likelihood so completely strange that we could not even begin to count the ways they may be different—and the problem looks insoluble.

I don't believe things are quite so bad. We needn't guess every facet of a visitor's behavior; we only need to check if they have left anything behind as a calling card. Such an artifact must be recognizably artificial. If it was constructed by an alien intelligence it should stand out—indeed, was designed to do just that. Even here there are limits, though.

Let's look, for a moment, at one of the natural world's most fascinating coincidences. At total eclipse, when our moon moves between us and the Sun, a beautiful display occurs. The round moon perfectly overlaps the Sun's disk. Streamers of hot gas continually boil up from the Sun's surface, jets forming a bright halo around the moon. This is a remarkable effect, quite dramatic. No other moon in the solar system provides such a floor show for its planet.

Is this accidental? The moon has not always been in its present orbit, after all; tides drive it away from Earth at a rate of about two feet a year. A few hundred thousand years in the past (or the future) this exact overlapping of the Sun's and moon's disks would not occur. No doubt eclipses would still be impressive, but much less so than now. It is striking that man's inquiring intelligence has evolved at just the right moment to appreciate this beautiful accident.

Or perhaps this additional coincidence—man's rise to civilization and the exact eclipse occurring at the same time—is no coincidence at all. Did some alien visitor accurately predict the evolution of intelligent apes and leave this massive signpost in our sky? The idea may seem absurd, but it is not impossible. This notion illustrates that the boundary between artifacts and natural coincidences is blurred.

Still, though we cannot anticipate the possible accomplishments of an alien technology, we can require that our knowledge of physical law and information not be violated.

Let's take a simple example. There is a cliche story plot that runs something like this: An alien interstellar expedition runs into trouble and must make an emergency landing on a lush, green, uninhabited planet. There is another disaster; maybe the native animals or bacteria kill most of the ship's crew. All but two of the aliens die. In the closing lines we learn that the virgin planet is third from its star, and the two aliens are named Adam and Eve.

This is nonsense for several reasons. The alien body chemistry would have to be exactly like the native life's—digesting the same sugars and amino acids, manufacturing blood cells based on hemoglobin, requiring precisely the same vitamins, and so on. This is most unlikely.

Even worse, how could we then explain the similarities we have with other primates? Did aliens bring them along, too? Fossil evidence shows a clear line of descent for mankind, all the way from twenty-five million years ago. Our ancestor Proconsul certainly wasn't smart enough to build a spaceship, and even if he had been, why was he so peculiarly adapted to the ecology of Earth? Similarly, the integrated nature of all Earthly life, with its common chemical schemes and DNA-based reproduction, strongly suggests that no alien biology ever gained a foothold here. No, we are undeniably the sons and daughters of this Earth. We weren't dropped into our niche by accident. Leaving behind wildly un-Earth-like organisms as a signpost would be a tricky proposition, simply because they would have to compete with hardy natives (not to mention the problem of differing metabolism discussed above). Recent discoveries of plants using unusual chemistries in deep-sea volcanic vents prompted some speculation about how ancient these forms were, but they do seem to fit into local evolution. Unless even stranger forms lurk in some dark corner, we may dismiss the idea of bio-artifacts as calling cards.

In the same way, we have to be careful about accepting any historical evidence of extraterrestrials. We need something more than vague legends about marvelous, miracle-working beings who live in the sky. Virtually all religions, past and present, require that the gods live underground and/or above the clouds. (After all, where else could they live—over the next hill? Then an unbeliever could refute an entire theology in an afternoon's walk.) Recently Soviet ethnologists conjectured that stories from the Bible are garbled versions of alien visits—even that Sodom and Gomorrah were destroyed by an atom bomb. There is simply no evidence for these ideas.

Enough time has passed to erase radioactive elements or any other signature. Similarly, biblical accounts of spaceship-like objects in the sky leave us
with nothing to check.

A legend is only an aged yarn leading nowhere. We need something solid
and unmistakable before such theories become anything more than arm-
chair speculation.

At various times people have come forward with artifacts that they
thought were evidence of alien visits. Some were well-meaning and others
outright frauds. A Scotsman thought fused towers in Ireland and Scotland
were works of high technology; it turned out they had been fired with peat,
a process the Scotsman didn't know. Etruscan gems were mistakenly taken
by some to be gifts from extraterrestrials because they seemed strange and
sophisticated; they were finally proven to be man-made. Dr. Gurlt's Cube, a
steel parallelapiped found imbedded in an ancient bed of coal, got quite a
lot of press coverage but proved to be a hoax.

Lacking any historical evidence we can be reasonably sure of, we can
turn our attention to the possibility that other beings left artifacts on or
near Earth even before (or while) man evolved. In *2001: A Space Odyssey,*
Arthur C. Clarke and Stanley Kubrick imagined such an artifact to be a
monolith and endowed it with a decidedly theological purpose: the uplift-
ing of man. This was a good dramatic device, but it is certainly not the only
role an artifact could fulfill.

A second function of the monolith was set forth in the source for *2001,*
an Arthur C. Clarke story, "The Sentinel." Here the idea was that the aliens
left some guidepost or sentinel that would trigger a signal (a warning?)
when man reached a certain technological level. This seems reasonable if
our visitors wanted to know immediately when we developed. After all,
they might want ample notice that we had discovered nuclear weapons or
space travel, rather than learn a few centuries hence, when we come
visiting them.

There is another role an artifact can play, beyond that of evolutionary
guide or sentinel. Keep in mind that the distances between the stars are
vast, our galaxy is many billions of years old, and intelligent races may not
live very long in comparison. An alien expedition passing through our solar
system long ago might never expect to return, or even to remain in this
neighborhood of the Galaxy. They may, from experience with other races,
know that civilizations do not survive for long on the cosmic time scale.
Chances that Earth would evolve a culture worth knowing while our
visitors' society was functioning were quite remote.

**Figure 1.** The most famous alien artifact in science fiction is the Monolith from Clarke and Kubrick's *2001: A Space Odyssey.* Here, it is discovered on the Earth's moon. *(Photo: Copyright MGM 1968.)*

So perhaps they would leave some sign on Earth, as if to say, "Your intelligence is not alone." Even more, their artifact might serve as a legacy. In it they could store information, both scientific and cultural, which could extend the lifetime of a civilization. Perhaps it would serve as an epitaph for their own race, a kind of last defiant gesture against the forces of entropy that could bring down intelligent societies. On a small chip we can even now write an enormous wealth of knowledge; not much space would be required to leave a rich library inside some relatively indestructible vault.

Much cannot be conveyed by simple language alone, as every artist will tell you, and so objects might be left in a vault as well. Such a legacy would be of unimaginable benefit to mankind, a sort of colossal Pharoah's tomb containing new science and new culture. Discovery of this legacy would be the most important event in human history.

Monolith, sentinel, legacy—these are guesses, necessarily anthropomorphic ones, at the motivations of superior beings.

Then too, the agency that leaves the artifact may not be a living member of another race at all. The stars might be explored by computerized ships, not flesh and blood.

Sending living beings on voyages many light-years in length is expensive and inefficient, compared to computer-directed flights. NASA's experience indicates that unmanned probes cost a thousandth as much as manned ones. Even a vast, wealthy society would probably prefer to send unmanned ships for exploration and use members of their own species only on definite missions (say, colonization) to a known destination. John von Neumann, a great, innovative physicist, imagined that advanced computing machines could self-reproduce, using only raw materials. Set loose in the Galaxy, such machines could spread like rabbits in Australia. If so, we should see a night sky clogged with orbiting craft looking for asteroids to devour. We don't, so probably this means that intelligent space-farers don't unleash galaxy-gobbling devices, for sound, "ecological" reasons. After all, they may well value contact with others for cultural exchange, and see more in our solar system than a large asteroid belt to mine.

Even if we dismiss the argument that earlier alien societies would have filled the Galaxy, however, we should also remember the vast time scales that figure into interstellar exploration. Light takes many years to travel from one star to the next. Spaceships, even if they can travel at near light speed (a feat we are nowhere near mastering), would take very long indeed to visit the hundred billion stars of our galaxy. Man has been around in identifiable form only about two million years. Cro-Magnon man (ourselves) is about thirty-five thousand years old. Written records go back seven thousand years. These may seem like long times to us, who are granted only three score and ten. But life has been on Earth for billions of years and could have aroused the curiosity of a passing computerized spaceship long before man appeared.

If intelligence inevitably arises, once a stable ecology evolves on a planet, automated visitors might have been under instructions to leave some artifact. Where would they put it?

One obviously safe place is away from Earth entirely, in orbit about the Sun. A billion years ago it might have seemed equally likely that intelligence would evolve on Mars *or* Earth, so a signifier might be left in orbit between them. This probe, left behind by the main ship, would wait—drawing power from the Sun—until it detected signs of intelligent life on a nearby planet. From such a distance the only reliable evidence would be radio signals, the first indication of what we call modern technology.

If a radio signal ever came, the probe could simply repeat this signal, aiming its radio beam back at the source. A computer program could establish a common language, once firm contact occurred. This system has the advantage that the probe's radio signal, coming from a nearby orbit, would be much more powerful than a beam from the probe's home star, many light-years away. Also, the target planet (Earth) need not have very sensitive receivers.

Such a repeated playback would undoubtedly attract the attention of the natives—imagine the surprise of Marconi if he had found a mysterious echo to every transmission he made. For our purposes, the matter might be laid to rest right there: Marconi heard no echoes, therefore such a probe doesn't exist.

In 1935 Carl Stormer and Balthasar van der Pol, studying the atmospheric propagation of radio waves, detected echoes returning many seconds after the original signal. The time lag indicated reflection from an object many times as far away as the moon. These stood as riddles until the 1970s, when a British electronics engineer, Anthony Lawton, repeated the experiments. He showed that refraction of radio waves in the ionized regions of our upper atmosphere can give such delays, and concluded wryly that "long-delayed echoes would be a most cumbersome and unnecessarily confusing way of making contact. Surely the obvious thing a probe would be programmed to do would be to send its *own* signals and make itself as conspicuous as possible."

It seems unlikely that a probe orbiting around our Sun would simply stay there, giving no sign of its presence and waiting to be found. Picking out even a mile-sized object so far away is very difficult.

There are ways of narrowing down the search, though. In the early 1980s F. Valdes and R. A. Freitas, Jr., searched the lunar Lagrange points for artifacts. These are locations in space that allow remarkably stable orbits. An object left there would not be tugged by the weak but persistent influences of the distant planets. There are two Lagrange points near our moon where the combined influence of the Earth and moon would allow an object to remain fixed for roughly a billion years. Valdes and Freitas looked for shiny objects reflecting sunlight and found nothing at these points, down to their observing limit of objects a few meters in size. (This was reported in *Icarus,* Vol. 55, p. 453.)

In the middle 1980s Michael Papagiannis of Boston University suggested using the infrared images of the night sky in another, similar pursuit. Thus far Papagiannis has not found enough funding from NASA to conduct his search.

Without such a neat signature, there is little reason to visit just any small dab of light that orbits our Sun. One place we have visited at great cost, though, and will probably go on visiting, is our moon. Aliens might well have put the artifact there—whether monolith, sentinel, legacy, or some unimaginable variant. It would be difficult to ensure the stability of orbits around the Sun or Earth for millions of years (except at LaGrange points), but planting the artifact on the moon would anchor it securely.

Covering the artifact with a few feet of lunar dust would bring another benefit: no more erosion by particles streaming out of the Sun (the solar wind) or interstellar space (cosmic rays). True, by the same stroke it becomes vulnerable to the occasional selenological activity on the moon. Also, large incoming meteorites could still damage it. These are unavoidable dangers, but by leaving several widely spaced artifacts scattered over the moon, the odds against all of them being destroyed by outsized meteorites can be made quite good.

There is another trouble with this theory—we have circled the moon and landed on it, and no friendly radio message came out to greet us. Should we conclude that nothing is waiting for us there? Not necessarily.

For one thing, Kubrick and Clarke could be right. The tip-off that a monolith lies buried somewhere may be subtle, such as a local warping of the magnetic field. Or, as in "The Sentinel," the object may be a small pyramid sitting atop a mountain peak that literally must be stumbled across before humans recognize it for what it is. An artifact should try to attract attention, so it should be large and bright. But surface damage could have eroded these eye-catchers in a billion years, so we may need to look for rather subtle features. If these ideas are right, only a full-scale exploration of our moon would turn up anything interesting.

On the other hand, the radiation damage mentioned above may be quite important. Sensitive electronic components made today cannot withstand constant bombardment by high-energy particles; they must be shielded. If an alien artifact were to remain on the moon, operating for millions or even billions of years, blocking out the radiation must be a very serious issue. This means the object will be buried at least a few meters deep. Moon dust and gravel above it will stop particles, yes—and also radio transmissions.

To get around this, it would probably be best to have the artifact periodically protrude a radio antenna to the surface. There it would listen for transmissions from the Earth. The Sun is somewhat noisy in the radio wavelengths. To cut down on this noise, a sophisticated artifact would probably surface its antenna when the Sun is below the moon's horizon.

What is a reasonable interval between appearances of the antenna? There is absolutely no way to tell. Over the long wearing course of eons, even a simple matter of extending an antenna can run afoul of accidents, so the period should not be too short. On the other hand, if the artifact were left in the first place to keep track of how rapidly humanity advanced, a frequency of once a century might be reasonable; this is roughly the life span of the beings that evolved on a planet, and may represent the minimum time needed to bring about a major technological advance. A sentinel that last emerged in 1890, before Marconi, would then be readying to look again in 1990.

All these factors depend strongly on precisely why the artifact was left. If it were a technological sentinel, to announce our graduation to the space-flight level, the moon is a good site. Even better, place it on the moon's other side, away from Earth. Then no freak radio contact with an Earth-bound station would have occurred in our history.

If something is buried on the moon's far side, and it does not surface frequently, contact with it may be postponed indefinitely. NASA plans (and probably Soviet ones, too) call for very little activity on the far side in the foreseeable future—and difficulty of direct radio communication is one of the reasons. We may be in for a long wait for a call from a far-side sentinel. Nonetheless, it would be wise to keep an ear cocked for a stray signal that might be an aged but still functioning near-side sentinel, struggling to get through our ionosphere and be picked out of the commercially generated noise we ourselves are making. It would be ironic if we were blotting out word from the stars with a thick layer of cornflakes advertisements and *Star Trek* episodes.

But our visitors may be very wise beings indeed, and realize that all cultures need not develop very far technologically. What if Earth's natives never reached orbit or the moon?

Lacking any data but out own case, we have no idea how probable it is that intelligence and technology are linked. Certainly if Earth were a planet of an older star, our crust would have fewer metals and heavy elements and we would have a hard time building spaceships. More to the point, do intelligent creatures necessarily desire technology? Our visitors must— otherwise they would never get here—but they might have encountered races that simply didn't think along technological lines.

We may have an example of such a race already on Earth: the dolphins. Our descendants may well remember our inability to recognize dolphin intelligence as our greatest folly, because we do habitually equate thinking with tool making. But an alien visiting Earth two million years ago might have found the dolphins the obvious evolutionary path for high intelligence. Dolphins are conspicuous in the oceans, making acoustic gossip picked up miles away, but who would take the time to scour the African forests for elusive tribes of tool-using primates? Or, realizing that dolphins and primates (which are the same age, evolutionarily) both had a good chance to form civilizations, aliens might have decided to leave a legacy that would be reached by both species.

After all, if you're leaving a legacy for a race you will very probably never meet again, does it matter whether they are fish or land-rover, tool users or not? The dolphins might never discover fire, develop chemical fuels, or alloy metals—they certainly haven't yet. So they couldn't reach the moon, even though they might have great use for the cultural record left by the visitors. Exploring the land would be difficult for dolphins, and flying in the air more so. The most obvious spot to leave a legacy for the dolphins would be the oceans.

The trouble with leaving any artifact beneath or near the wind, wave, and tide of the sea is obvious—corrosion. The aliens would need to be sure their legacy would be read and understood very soon—in which case, why not just teach it directly to the dolphins during their visit? (Which raises an interesting possibility: Perhaps the dolphins already have the legacy, transmit it by word of mouth to each generation, and don't consider us worthy of receiving it. Touché, fishermen!) In any case, an artifact left in the sea is surely gone by now; corrosion is swift. There remains the possibility that the legacy might be left on land, either for us for some future dolphin civilization.

Where would it be? Some place with little erosion, certainly, far from the oceans, away from areas of geological activity or places where large land animals could interfere with it. If the drifting of Earth's continents is typical of planets, and if our visitors knew the dynamics of plate tectonics, they could have selected sites with few earthquakes, volcanoes, or other severe changes.

The erosion rate is high near mountains and glaciers, so we can write off the great mountain ranges and many sites too near the North or South Poles. The great band of tectonic stress that loops over our planet like a baseball seam makes many other places, such as the California coast, unlikely. The Canadian sheet, a great area of very old, stable formations, would serve quite well if the glaciers had not so (relatively) lately steamrollered it. And so it goes for a large fraction of Earth.

Two sites do look promising: the interior desert of Australia and some southern portions of the Mongolian plateau. Australia offers the added bonus of being relatively cut off from Africa, where man apparently evolved. Our visitors might well have decided to leave any artifact as far from Africa as possible, reasoning that we would be further advanced by the time we found it.

Both these sites are relatively unexplored even today. Little grows there and few animals of any size are native. Many parts of Australia in particular are extremely hard to reach without mechanized transport or great endurance.

I am not a geologist, and the matter of guessing what sites have been most stable for very long times is a complicated one, best left to experts. The important point is that such sites may exist. Just this knowledge is not enough, though, because we return to the essential mystery of this whole discussion: Who (or what) are we dealing with? What sort of artifact would be left behind? How can we recognize it?

Obviously it should not seem natural. But after lying on Earth's surface for perhaps millions of years, it might not look very artificial by now. It must be covered by dust and gravel at least, if not rock formations of much greater weight. Without knowing specifically what its builders had in mind we cannot reasonably guess whether it could be designed to stay above-ground or not. It is impossible to say whether aliens who can fly between the stars (or send computerized ships instead) possess materials that can resist normal erosion, or have other special properties that give them away. Certainly if the artifact were large enough—say, the size of a mountain— and very regular in shape, we could spot it easily enough. Since we haven't, we should look for some more subtle beacon.

There are many attention-getting signs that do no depend on size—regular arrays, say, or differences in the quality of the light an object gives off (polarization or odd spectra). What kind of regularities? What sort of light? The possibilities are endless. For example, radar searches from orbit have already peered beneath the Sahara sands, finding "fossil" river valleys in the rock layers below. Radial rays cut in rock, pointing to a central origin, would be a simple sign of a legacy, safely buried under desert. Other methods will doubtless emerge as our surveys improve. The best we can do is look for the unexpected, and look very thoroughly.

Here we are in luck. We have detailed aerial photographic surveys of the our planet, made both from orbit and low-flying aircraft. Wide regions never before studied have been photographed through infrared and ultraviolet filters. The photographs are well systematized, so that information can be found from them conveniently and efficiently.

In the 1970s, after the first photos from Mars arrived via Mariner, it was pointed out that similar pictures of Earth with an optical resolution of one kilometer would have shown little evidence of man's civilization. Certainly we will need far greater resolution to see even one lone artifact standing in an arid desert. Thus, photos that display great detail will have to be painstakingly analyzed with a completely open mind.

A curious arrangement of a ridge line, a concentric pattern of rock formations, perhaps an abnormally high reflectivity in the ultraviolet or odd polarization—any of these could be either pure accident or, on the other hand, the first subtle clue.

Admittedly, there may be nothing in the Australian Outback or anywhere else. But a survey designed to discover such sites—or oddities on the moon, in the asteroids, or at spots like the Lagrange points—could yield interesting and significant insights into conventional astronomy or geology.

Healthy skepticism is important. We risk being discredited if we give each possibility less than scrupulous study. For example, the "face on Mars" ferreted out from space probe photos of that surface is certainly a strange formation, and may repay study. It is almost certainly a natural effect, but overenthusiasm by cranks for its possible ETI origin has already driven many scientists away from further work on it.

Passive listening for radio calls requires that aliens send just when and how we choose to hear. The chances of this temporal coincidence may be quite small. Searching for artifacts can succeed if visitors passed by any time in the last several billion years.

This advantage should encourage us to undertake inexpensive searches. Computer analysis and image enhancement of Earth surface photos promise to make such studies possible as side projects to more mundane surveys, using few resources. The important point is that flexibility and an eye for oddities may yield enormous, incalculable rewards.

# CHAPTER 5

# MEGA-SETI

*SETI is based on the fundamental fact that radio waves travel at the speed of light, while spacecraft will necessarily be much slower.*

*The Universe is vast. The distance across our Milky Way Galaxy is some 100,000 light-years. Other galaxies are millions, even billions of light-years away. The fact that we can see the light of their stars in the sky means that we could also receive their radio signals, if they were sent with enough power. If they were sent at all. If there is anyone out there to send signals.*

*In this chapter, Michael Papagiannis, first president of the new I.A.U. Commision 51 on Bioastronomy shows how radio searches of the sky began, and who is doing them today. Most of those who have given SETI any serious thought agree that radio searches are the best technique we have to make contact with extraterrestrial intelligence. That is why continued support of these efforts is so vitally important.*

# THE HUNT IS ON: SETI IN ACTION

## BY MICHAEL D. PAPAGIANNIS

### THE DAWN OF RADIO SEARCHES

**T**hroughout history, schemes proposed or used to contact extraterrestrials have reflected the state of science of the times. With the Industrial Revolution of the nineteenth century, we begin to see more realistic ideas about how to contact other civilizations that were thought to inhabit other parts of our solar system, but none of them was implemented. The modern era of actual searches for extraterrestrial life and intelligence finally began in the twentieth century, following the development in the 1950s of the field of radio astronomy and the advent of the Space Age.

The first cosmic radio signals (radio noise from the center of our galaxy in the direction of Sagittarius) were detected by Karl Jansky in 1932, while he was studying radio interference from terrestrial sources at the Bell Telephone Laboratories in New Jersey. In 1938, Grote Reber built, in the backyard of his home in Wheaton, Illinois, the first intentional radio telescope. It had a dish thirty-one feet in diameter, which was used to produce the first radio map of the sky at a frequency of 160 MHz, near today's TV frequencies. Jansky showed that the galactic center, and a few discrete sources in the plane of the Milky Way, were strong sources of continuous radio static at several frequencies, and that the Sun was a weak radio source (except at times of solar flares).

During the Second World War, because of military needs (radar, etc.), there was great progress in radio equipment. A number of people observed the Galaxy's radio output at various frequencies and directions, always finding it to be of roughly uniform brightness independent of frequency. In 1951, Professor Edward Purcell of Harvard and his graduate student, H. Ewen, by looking along the Milky Way and observing a modest band of frequencies, managed to detect the first radio spectral line, a bright spike in the general static, emitted by atomic hydrogen centered at 1.42-GHz (at a wavelength of twenty-one centimeters), which had been predicted during the war by the Dutch astronomer Van de Hulst. Then in 1959, in their famous paper published in *Nature,* G. Cocconi and Philip Morrison recommended searching for radio signals from other stellar civilizations, and doing it near the hydrogen line frequency, which, they wrote, "Is a frequency that must be known to every observer in the Universe." They closed their paper with the statement, "The probability of success is difficult to estimate, but if we never search the chance of success is zero."

Only months later, in the spring of 1960, Frank Drake used the then-new 26 meter (85-foot) Tatel radio telescope of the newly established National Radio Astronomy Observatory (NRAO) in West Virginia to conduct the first radio search: looking for ETI signals near the hydrogen line frequency from two nearby Sun-like stars—Epsilon Eridani (10.2 light-years) and Tau Ceti (11.9 light-years). He observed them for about two hundred hours in a project he named Ozma (after the princess in *The Wizard of Oz*). Project Ozma marks the birth of modern SETI, i.e., of the new field of science that pursues "the search for extraterrestrial intelligence."

Looking back, we see that in just three decades of space exploration, we have landed men six times on the moon, and landed more than a dozen probes on Mars and on Venus; we have flown missions by the planets Mercury, Jupiter, Saturn, Uranus, and Neptune, and the space probe Pioneer 10 has already gone beyond all of the planets of our solar system. Soon it will become lost in the Galaxy, carrying on it a golden plaque with the figures of a man and a woman, like a greeting card from the people of Earth to any extraterrestrial civilization that may find it.

The searches for life in our solar system have so far been negative, including the two Viking probes that landed on Mars in 1976 and tested its soil for biological activity and for the presence of organic compounds. Probes, possibly some manned ones, will hopefully be sent to Mars again during the next ten to twenty years to study our neighboring planet more carefully, because close-up photographs indicate that Mars may have had periods with liquid water on it. Probes are also being planned for Titan (the Cassini mission), the large moon of Saturn where prebiotic chemistry seems to be active, and for Europa (the Galileo mission), the second of the

**Figure 1.** Karl Jansky and his rotatable radio antenna. With it he discovered that radio noise is emitted by celestial sources. *(Photo: Courtesy AT&T Archives.)*

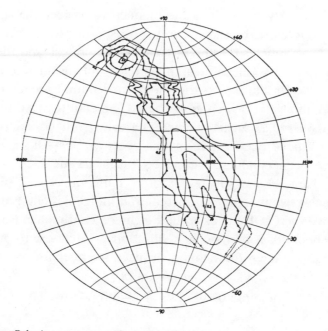

**Figure 2.** Grote Reber's contour map of celestial radio noise. The main contours coincide with the outline of the Milky Way and the galactic center. The concentrations at the upper left were eventually identified as a supernova remnant, and colliding galaxies. *(Source: The Milky Way.)*

Galilean moons of Jupiter, which under its icy crust may have a subterranean ocean of liquid water and hence the possibility of life. Over all, however, the probability of finding life anywhere else in our solar system seems at present to be quite low.

The search for direct evidence of *primitive* life in other solar systems is still beyond our technical capabilities. With the rapid development of space astronomy, though, it seems likely that in the next ten to twenty years we will have large astronomical observatories in space and/or on the moon. These will be free from the problems caused by Earth's atmosphere and will allow us to make a more stringent search for planets around other stars than we have been able to carry out from the ground. Ultimately we hope to discover Earth-like planets, and to study them spectroscopically, searching for spectral lines of oxygen and ozone that would signal the presence of life. (Oxygen and ozone are highly reactive chemically, and therefore they cannot long exist in a planetary atmosphere without being continuously replenished through biological processes similar to photosynthesis.)

At present, however, the most likely way to detect the presence of life elsewhere in the Universe seems to be through radio signals sent to us by advanced civilizations. For this reason, SETI is now the area where we are focusing most of our search efforts.

## THE DEVELOPMENT OF SETI

In the nearly thirty years since the pioneering Project Ozma, scientists have carried out close to sixty search projects, most of them at radio frequencies, but a few also in the ultraviolet, the optical, and the infrared. Jill Tarter of the NASA-Ames Research center maintains an active file of all these projects. So far all projects together have accumulated approximately two hundred thousand search hours, and ten countries (United States, Soviet Union, Australia, Canada, France, Germany, Holland, England, Japan, and Argentina) have been active in this effort. Almost all of these radio searches have been conducted at select "magic frequencies" as they are called—the characteristic spectral lines of common atoms, such as the hydrogen line at twenty-one centimeters, and molecules and radicals, such as the four hydroxyl (OH) lines around eighteen centimeters (about 1.665 GHz), in the hope that other stellar civilizations will choose to broadcast at these universally known frequencies to make it easier for young civilizations, such as ours, to pick up their messages.

Probably more than 90 percent of the search hours we have accumulated

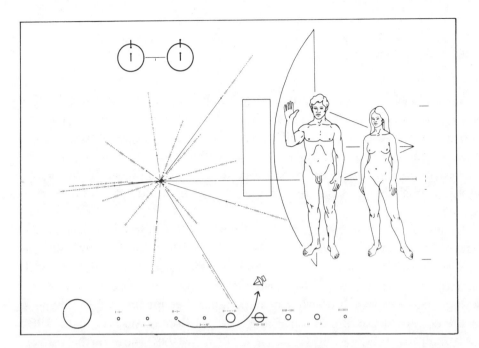

**Figure 3.** An image from the plaque being carried by the Pioneer spacecraft to the stars. It bears a schematic of the Pioneer's path out of the solar system, the hydrogen molecule, and the directions from Earth to various pulsars—along with codes for their frequencies. This information would give alien astronomers the date and place of origin of the spacecraft. *(Photo: Courtesy of The Astronomical Society of the Pacific; NASA.)*

**Figure 4.** Pioneers of SETI. From left to right: Michael Papagiannis, Philip Morrison, and Frank Drake. *(Photo: Courtesy Michael Papagiannis.)*

so far have been at the hydrogen line frequency, but unfortunately this strategy has not paid off. As a result we are beginning now to reorient our efforts toward searches that will cover much wider ranges of frequencies. This was not possible in the early days of SETI, but technology has advanced rapidly in these thirty years, and we can now search for signals simultaneously in many different frequencies using Multi-Channel Spectrum Analyzers.

An important step in the development of SETI was the fact that the International Astronomical Union (IAU), the world body in astronomy, in 1982 espoused this new effort and established a special commission for it, named, IAU Commission 51—Bioastronomy. The term "bioastronomy" describes the new branch of astronomy that searches for life and intelligence in the cosmos. I was the first president of IAU Commission 51, from 1982 to 1985. I was followed by Frank Drake (1985–1988), now of the University of California, Santa Cruz, and by G. Marx (1988–1991), of Eotvos University in Hungary. IAU Commission 51 grew rapidly to a membership of three hundred astronomers from thirty-three countries, established its own newsletter (*Bioastronomy News*), and held its first IAU Symposium in 1984 in Boston, its second in 1987 in Hungary, and will hold its third in 1990 in France. Following the success of our first IAU Symposium, the science magazine *Nature* asked me to write a review article, which was published in 1985. The success of this new IAU commission was a clear indication of the strong interest that exists for this field within the scientific community, as well as among the general public.

Two other important contributions to the acceptability of SETI also occurred in 1982. One was a letter in support of SETI that was prepared by Carl Sagan, signed by seventy prominent scientists, including eight Nobel laureates, and published in *Science*. The other was a recommendation from the committee on astronomy of the U.S. National Academy of Science for the expenditure of funds to advance SETI in the 1980s.

The SETI projects conducted so far can be divided into three general groups: directed; shared or parasitic; and dedicated.

## DIRECTED SEARCHES

These searches use a major observatory for a relatively sort period to carry out a specific SETI project. Typical examples are the following:

P. Palmer and B. Zuckerman used the 91-meter (three-hundred-foot) antenna of the National Radio Astronomy Observatory (NRAO) in West

Virginia to study the hydrogen line frequency looking toward 674 stars, in five hundred hours of observations. This project, which they named Ozma II, was conducted intermittently over a four-year period between 1972 and 1976.

J. Tarter, et al., from 1979 to 1981, used the Arecibo radio telescope in Puerto Rico (which, with a diameter of 305 meters, is the largest in the world) to observe 210 Sun-like stars in a 4-MHz band around the magic frequencies of hydrogen and hydroxyl. They also used a new technique and a CDC 7600 computer to obtain an excellent frequency resolution of 5.5 Hz.

Lord and O'Dea used the fourteen-meter (45-foot) diameter millimeter-wavelength radio telescope of the University of Massachusetts to look near the frequency of the 115 GHz line of the carbon monoxide molecule for strong beacons along the north portion of the rotational axis of our galaxy. This is a potentially magic location for beacons operated by supercivilizations, because from such a vantage position they would be able to have direct line-of-sight access to every part of the galactic disk, unobstructed by the dust and gas that lies in the plane of the galactic disk.

Another interesting search was undertaken in 1983 by Freitas and Valdez, who used the Hat Creek radio telescope in California to conduct a search at the 1.516 GHz line of tritium, which is the heaviest isotope of hydrogen (one proton and two neutrons). Since tritium is unstable and has a half-life of only 12.5 years, if found in the vicinity of normal stars it must be of artificial origin and hence most likely the by-product of huge nuclear fusion plants.

None of these searches had positive results.

Freitas and Valdez also used the seventy-five centimeter (thirty-inch) optical telescope of the Kitt Peak National Observatory (KPNO) to search for artifacts and probes possibly placed in our solar system by extrasolar civilizations. They looked for them at the magic locations of the L4 and L5 Lagrange points of the Earth-moon and Earth-Sun systems. These are regions in space where objects tend to remain, even if perturbed by outside forces, and, therefore, the regions from which extraterrestrial stations could be observing Earth. Five of these special points were discovered by the French mathematician and astronomer Louis Lagrange (1736–1813). Three of them (the less stable) are along the line that connects the two major bodies, while each of the other two (L4 and L5), which are much more stable, form equilateral triangles with the two major bodies and travel with the smaller of the two, 60 degrees ahead of it and 60 degrees behind it, in its orbit around the larger one.

In 1978, W. Sullivan, et al., of the University of Washington, proposed that an alternative search strategy for ETI beacons would be to look in the direction of nearby stars for radio signals leaking unintentionally into space from advanced civilizations. As a test, Sullivan and Knowles used the Arecibo radio telescope to observe the radio leakage of Earth in the 150–500 MHz range, by observing reflections from the moon. Strong television stations and powerful military radars were the most prominent sources seen. The space surveillance radar of the U.S. Navy in Archer City, Texas, for example, which operates at 217 MHz, emitting pulses in all directions with total power of fourteen billion watts in a bandwidth of only 0.1 Hz, could be detected by a civilization with Arecibo-type telescope technology up to a distance of *twenty light-years.* Knowles and Sullivan also did a limited eavesdropping search on a few nearby stars, using the Arecibo radio telescope and the Mark I VLBI (very long baseline interferometry) system, which, because of the large distances that separate its various radio telescopes, is able to achieve a high angular resolution and thus focus on a small region of the sky. Although this test run found no signals, it did show that eavesdropping is an alternative search process that is feasible and can also be pursued.

## SHARED AND PARASITIC SEARCHES

Shared and parasitic searches either reanalyze archived data that had been obtained for other purposes for ETI signals, or they share in a parasitic or piggyback fashion the data being obtained by a radio telescope for an unrelated project, and process them for ETI signals as they are being obtained.

F. Israel of the Netherlands, working in 1981 first with DeRuiter and then with Jill Tarter, reanalyzed many of the 21 centimeter "noisy" sky maps that had been obtained with the Westerbork telescope array of Holland and had been stored by the Dutch astronomers. They were looking for strong radio sources in positions that coincided with known stars. In another project N. Cohen, et al., in 1980 reanalyzed their own surveys of globular clusters, looking for narrow-band signals at the frequencies of OH and $H_2O$ masers (pure-frequency, high-intensity beams from dust clouds), at 18 and 1.35 centimeters respectively, that may get tapped by supercivilizations to produce strong signals in certain directions.

In 1980, S. Bowyer and D. Werthimer of the University of California, Berkeley, and their coworkers built an automated 100-channel spectrum analyzer to siphon off data obtained by a radio telescope for an unrelated project. This parasitic or piggyback SETI device was named SERENDIP and was used with the Hat Creek and Goldstone antennas. It was later upgraded to SERENDIP II, which has a 131,072-channel fast Fourier processor with a resolution of 0.49 Hz per channel. The system can operate unattended on a twenty-four-hour basis with practically any radio telescope.

In the mid-1980s I undertook a search for large artificial objects (such as space colonies or processing plants) in our solar system, and especially in the asteroid belt, which is an ideal source of raw materials (metals and water), to test the theory of galactic colonization; that is, the possibility that the entire Galaxy might have already been colonized by supercivilizations. I used the infrared astronomy satellite (IRAS) data bank of infrared sources in our solar system (mostly asteroids and a few comets). IRAS was developed by a group of scientists, of which I was a member, who assembled at the Infrared Processing and Analysis Center of Caltech to look for infrared objects in our solar system. My intent was to scrutinize about ten thousand images for infrared spectra that were too hot to be due merely to reflected sunlight, in the hope that artificial materials of energy-intensive industry would show up, much as our cities stand out as "heat islands." Unfortunately, due to insufficient funding I was not able to complete my search.

## DEDICATED SEARCHES

Dedicated searches are performed by radio observatories that have SETI-dedicated facilities and conduct searches on a continuous basis. Two such projects now in operation account for most of the observing hours accumulated thus far. The oldest of the two is the Ohio SETI program, which has been in operation since 1973 under the direction of Dr. J. Kraus and Dr. R. Dixon. It uses the meridian-transit radio telescope of the Ohio State University, which has a collecting area the equivalent of a parabolic dish 53 meters (175 feet) in diameter. The search is conducted at the hydrogen line frequency with the help of a fifty-channel filter bank with a resolution of 10 KHz (kiloherz) per channel. The Ohio SETI program has been sustained with modest support from NASA and the tireless efforts of Kraus, Dixon, and many enthusiastic volunteers. A major false alarm in this project was the so-called "wow" signal recorded in 1977, which, in spite of repeated efforts, has not been found again.

Since March 1983 Professor Paul Horowitz of Harvard University and his collaborators have been operating the other SETI-dedicated facility using the 26-meter (84-foot) radio telescope of the Oak Ridge-Harvard-Smithsonian Observatory near Boston. This project is supported by the Planetary Society, a private organization headed by Professor Carl Sagan. Initally it had two 65,536-channel spectrum analyzers, with a frequency resolution of 0.03 Hz per channel. Horowitz and his colleagues then built a new MCSA with 8.4 million channels and a frequency resolution of 0.05 Hz per channel, which provides a total bandwidth of 420 KHz, enough to account for practically all the Doppler effects due to the relative motions of Earth and the transmitting source. This new system was named Project META and started operating in September 1985. It has already twice covered all of the available sky at the hydrogen 21-centimeter line and is now repeating the survey at twice the hydrogen frequency.

A third dedicated facility, also supported by the Planetary Society, will begin to operate around 1991 near La Plata, Argentina. It will be under the direction of Dr. Raul Colomb, and will be using the La Plata thirty centimeter (ninety-one foot) radio telescope. Two of the members of the La Plata team at Harvard worked with Horowitz to build the MCSA that will be used in Argentina, which will be the same kind as the one used at Harvard. The exciting aspect of this new SETI-dedicated facility is that it will be able to scan for the first time the southern skies that are inaccessible to the other two SETI-dedicated facilities, both of which are in the United States.

## THE NASA SETI PROJECT

The search for ETI radio signals is a multidimensional problem, which because of its complexity and size is often called the cosmic haystack. The dimensions of the "search space" include: coverage of the sky; frequency range; sensitivity; bandwidth; polarization; signal modulation; on-off periods; and more. Initially investigators, especially in the United States, had hoped that extraterrestrial civilizations would be transmitting at the hydrogen line to make contact easy. But extensive though not exhaustive searches by many observers at the hydrogen and a few other magic frequencies have produced no positive results.

With the rapid advance of technology, NASA is now getting ready to embark on the next generation of radio searches, namely to undertake systematic searches over wide frequency ranges in the microwave window (1–10 GHz) of Earth's atmosphere. This range is relatively free of cosmic radio noise and therefore is well suited for interstellar communications. Embedded deep in the microwave window is the "water hole" (1.4–1.7

GHz), which is especially noise-free, and therefore the most attractive range for interstellar radio communications. A point of great concern, however, is the rapidly growing use of this frequency range by our civilization for other purposes (such as satellite communications), which are bound to interfere with our future radio searches.

It is generally believed that SETI signals will be narrow-band to save power. Therefore, in order to have a good signal-to-noise ratio we need receivers with very narrow bandwidths, and since we want to explore a wide frequency range, we need a very large number of narrow-band channels. This function will be fulfilled by special MCSAs that NASA is now developing, each with 8.25 million channels (slightly fewer than META), but with a much wider total bandwidth than the MCSA of Horowitz. This new MCSA is now being constructed for NASA by a group at Stanford University headed by A. Peterson and I. Linscott. The highest frequency resolution will be about 1 Hz per channel, giving a total bandwidth of about 8 MHz. It will also be able to analyze simultaneously the incoming 8 MHz frequency band into channels of 32, 1,024, and 73,728 Hz.

It is obvious that there are going to be significant tradeoffs between the NASA SETI project and Horowitz's META project. The most important is the bandwidth of each channel, which in the META project is twenty times narrower and therefore can secure a much better signal-to-noise ratio for very-narrow-band signals. The NASA project, on the other hand, with channels that are twenty times wider, will be able to cover a much wider frequency range, allowing for almost any conceivable relative motion between us and a source, which is actually its main objective.

Signals that exceed a predetermined threshold chosen by the desired signal-to-noise ratio will be flagged for further testing. Sophisticated signal detection algorithms are also being developed by NASA, and a considerable effort is being made to achieve on-line processing of the data, a difficult task given the huge volume of incoming data with an 8-million-channel MCSA. Emphasis is also placed on the ability to detect pulsed signals and signals with a velocity-caused frequency drift (the velocity between us and the sender). It is expected that the first 8-million-channel MCSA will be ready in the early 1990s, but more realistically the whole system, including all signal-processing instrumentation, will become operational around the mid-'90s.

A prototype unit with 73,728 channels, together with several of the new signal-recognition algorithms, has already undergone tests with the sixty-four-meter (208 foot) diameter Goldstone antenna in California. In a recent test, it was able to pick up the very weak (1-watt) signal beamed by the Pioneer 10 spacecraft toward Earth, from a distance of about thirty-five astronomical units (five billion kilometers)—the fringes of our solar system. It was also able to see clearly the frequency drifts imposed on this signal by the rotation of Earth and the relative motion of Earth and the spacecraft.

The NASA SETI project is headed by Dr. John Billingham of NASA-Ames. His deputy is Dr. Bernard Oliver, and the project scientist is Dr. Jill Tarter; all are pioneers of SETI. The NASA SETI project will have two components. The targeted search will be the responsibility of NASA-Ames and, as mentioned elsewhere, will emphasize high sensitivity to weak signals by concentrating only on a number of discrete sources. The sky survey will be carried out by the NASA group at the Jet Propulsion Laboratory (JPL) headed by Dr. Michael Klein, and will scan the entire sky.

It is estimated that the NASA SETI project will take five to ten years, depending on how many of the 8-million-channel MCSAs become available. Thus, if it starts around 1992, we can expect it to be completed around the year 2000, or the year 2001 as someone with a cosmic sense of humor once said.

## PROGRAMS ABROAD

It is true that the United States has dominated the searches for extraterrestrial intelligence and has piled up many more search hours than any other country in the world. It is also true, however, that several other countries have made significant contributions to this effort. Most prominent among them is the Soviet Union, which was a leader in the early years of CETI (as they prefer to call it, using "Contact" instead of "Search" before ETI). In this effort, their leaders were I. Shklovskii, V. S. Troitskii, N. Kardashev, and V. I. Slysh.

Shklovskii wrote one of the earliest books on life in the Universe, published in the U.S.S.R. in 1963. A version in English, significantly expanded by Carl Sagan, was published in the U.S. in 1966 with the title *Intelligent Life in the Universe*. Shklovskii died in 1985, but in the last ten years of his life he had become pessimistic about finding extraterrestrial intelligence.

V. S. Troitskii was the leader in most of the searches that were conducted in the Soviet Union. In 1968 he carried out a search from his own Radiophysical Insitute in Gorki, targeted at ten nearby stars. In the 1970s he and Kardashev became the leaders of a major search that involved simultaneous observations from several stations around the Soviet Union, as well as from the ship *Academician Kurchatov* cruising near the equator. Measurements were made in the one-half to 10 GHz range, and the wide separation of the stations allowed them to estimate the direction of signals from the delays in their arrival at the different stations. It appears that they observed only terrestrial signals originating either in the ionosphere or in the magnetosphere of the Earth. In the early 1980s Troitskii was preparing a new search in Gorki that would have involved one hundred small (one-meter diameter) radio telescopes, operating at the hydrogen frequency, but he became seriously ill and the project was abandoned.

V. I. Slysh conducted a search at the hydroxyl frequency from 1970 to 1972 that was targeted at ten nearby stars using the Nancay radio telescope in France. More recently he conducted a sky survey of the 3-degree universal background radiation using a satellite radiometer at infrared wavelengths. The smoothness of the background ruled out the existence of Dyson Spheres, up to a distance of about three hundred light-years. Dyson Spheres are postulated megaconstruction projects, built by supercivilizations, which surround their star to utilize as much of its energy as possible, and thus their whole solar system would become a huge infrared (heat) source. A similar study was conducted in the late 1980s by Kardashev and Slysh using infrared data of the IRAS satellite to search again for Dyson Spheres (this seems to have become a preoccupation of the Soviet scientists). Their new sixty-five-meter (211 foot) millimeter-wave radio telescope in Samarkand is now approaching completion, and I was told that they plan to use it also for SETI.

Another interesting Soviet project, called Mania, was conducted by Victorij Shvartsman, who used the six-meter (19.5 foot) optical telescope of the Soviet Union in Zelenchuksksaya to search for optical pulses and for ultra-narrow spectral lines due to lasers. Unfortunately, Shvartsman died in the fall of 1987, and the project was discontinued.

Several search projects have also been conducted in other countries. In 1977, for example, Wielebinski and Seiradakis used the hundred-meter (325-foot) radio telescope in Bonn, Germany, combining a search for pulsars with a search for extraterrestrial signals. Starting in 1981 and continuing at intervals in the years that followed, Tarter and Biraud had been using the large radio telescope in Nancay, France, to observe three hundred stars at several of the magic frequencies. Paul Feldman and colleagues have repeatedly used the forty-eighty-meter (158 foot) Algonquin radio telescope in Ontario, Canada, to conduct radio searches, while in

1983 Sam Gulkis used the sixty-three-meter (205 foot) Tinbinbilla antenna of the NASA Deep Space Network in Australia to conduct a search of the southern skies at 8 GHz. I have already mentioned the SETI-dedicated facility that will be initiated in Argentina in 1991, and thus it must be quite evident that the search for extraterrestrial life and intelligence has finally become an international effort.

Characteristic of this development is also the fact that the "Flag of the Earth," which was designed by James W. Cadle and shows part of the yellow Sun, the blue disk of the Earth, and the small white disk of the moon against the black background of outer space, is now flying on all observatories that are doing SETI work, to show that this is a joint effort of all mankind.

## SOME CONCLUDING THOUGHTS

A simple comparison of the original Project Ozma to the Arecibo radio telescope with a powerful MCSA makes it obvious that we have made great technological progress in just three decades. A related question often asked then is why don't we postpone our searches until our technology becomes more advanced? I believe that the answer has two parts. The first is that we never know in advance the level of technology needed to succeed. It would have been a grave mistake, for example, to have asked the Wright brothers to wait for the discovery of the jet engine before trying to fly. They succeeded with far less, and on December 17, 1903, they opened the doors of the new field of aviation. I am sure that even if Drake could have known that twenty-five years later Horowitz would have an MCSA with 8 million channels for a search around the hydrogen line frequency, he still would have gone ahead with his Project Ozma, and rightfully so.

The second reason is that technology is like a ladder that we must climb one rung at a time, starting from the lowest ones. But as we climb higher, our horizons broaden and many new technological developments materialize—such as the jet engine in aviation and lightweight materials that withstand high temperatures in space programs, which also benefit many other fields.

After millennia of thinking and philosophizing about the plurality of worlds, we have finally entered the experimental era of SETI. We have already used space probes to search for primitive life in our solar system, and we are now searching with our radio, optical, and infrared telescopes for advanced civilizations in the Galaxy. It is a special privilege to live in the era that tries to answer experimentally profound, old questions about the prevalence of life, and especially of life with intelligence, in the Universe. With the NASA SETI project as the central force, and the many other parallel special searches now in progress or planned for the near future, we can expect that in the next ten to twenty years we will learn much more about the presence of other advanced civilizations in our galaxy.

If we are to find them, this would certainly be the greatest discovery in the history of mankind. But even if after concerted efforts we were to conclude that we must be one of very few if not the *only* advanced civilization in our galaxy, this too would be an important result. Because knowing how rare our civilization is, among the hundreds of billions of stars of our galaxy, would hopefully make us realize how cosmically important it is to preserve it.

# CHAPTER 6

# SETI ON CAMPUS

*Not all SETI programs require massive radio telescopes and equally massive government funding. Modest programs have been progressing quietly in Ohio, Massachusetts, and California. Even though they have not found that unambiguously clear signal from an alien civilization, they have helped to broaden our understanding of the stars and of the intricate technologies of radio communications and computer analyses of complex information.*

*Robert Dixon tells of the SETI program at Ohio State University, and how it was nearly wiped out by a real estate developer. Paul Horowitz and William Alschuler detail how a modest, privately funded effort led to the first dedicated radio-telescope search. Stuart Bowyer shows how SETI can be "piggybacked" on existing radio-telescope research programs.*

*These modest programs are amassing valuable data, including at least one signal that "wowed" the Ohio researchers and several unexplained spikes in the SERENDIP programs in California.*

*Have we already detected extraterrestrial intelligence?*

# THE OHIO STATE UNIVERSITY PROGRAM

## BY ROBERT S. DIXON

The Ohio SETI program got its first strong impetus from NASA's Project Cyclops. The goal of Cyclops, which was a paper study conducted in the 1970s, was to assess what it would take in terms of time, people, equipment, and money to mount a large search for radio signals from other civilizations. The end result was a report that was widely circulated as a NASA special publication, recommending a small array of radio telescopes, which would grow with time as needed.

During my Project Cyclops research, it became clear to me that many theoretical papers were being written about SETI, but nobody was actually doing any extensive searching. I also realized that we had a large, fully operational radio telescope available at Ohio State that was designed explicitly to search for new radio signals in the sky. (It had just completed the largest all-sky survey of natural radio signals up to that time.) Coincidentally, this telescope was also chosen by the Russian scientist Gindilis as the telescope most suited for SETI because of its unique surveying ability. Although we had no money, we had a crew of able volunteers on hand. Faced with the alternative of ultimately turning off the telescope and letting it rust away, we decided that we had a responsibility to seize the opportunity that had been thrust upon us and start a real SETI program. It did not take too much arguing to convince John Kraus, director of the OSU Radio Observatory, to allow me to use the telescope for the world's first full-time SETI program.

The Ohio State Radiotelescope is larger than three football fields in size, and equivalent in sensitivity to a circular dish 175 feet in diameter. The beam of the telescope is elliptical, being 40 minutes of arc in the declination (celestial latitude) direction, and 8 minutes of arc in the right ascension (celestial longitude) direction. This may be visualized by comparing it with the size of the moon, which is a circle 30 minutes of arc in diameter.

The telescope surveys the sky by remaining stationary and allowing the rotation of Earth to sweep its beam in a circular path through the sky once each day. After a few days of observation, the beam is moved slightly up or down, and the sweep is repeated. It takes several years to thoroughly search the sky visible from OSU.

We went on the air in 1973, using an eight-channel receiver system originally constructed for twenty-one-centimeter hydrogen line observations by Bill Brundage. He later became chief engineer of the 92-meter (three-hundred-foot) telescope at Green Bank, and was responsible for preparing the Very Large Array in Socorro, New Mexico, to receive Voyager spacecraft signals from Neptune.

The bandwidths of the channels ranged form 10 to 50 KHz, depending on their distance from the center frequency. The output of the eight channels was plotted as wiggly lines on pen recorders. The charts were laboriously searched by eye for unusual signals by graduate student Dennis Cole (now a contractor to JPL), and were used as the subject for his master's thesis in electrical engineering. This may have been the first graduate degree ever awarded in SETI.

The search strategy chosen then was to search in the vicinity of the 21-centimeter hydrogen line. Due to the random motions of the stars with respect to the Sun and their general revolution around the center of our galaxy, signals transmitted from another star system at the hydrogen line frequency (1.42 GHz) would be received at somewhat different frequencies because of the Doppler shift. To avoid this problem we made the deliberate assumption that any civilization transmitting at the hydrogen line frequency would offset their transmission frequency in just the right way to remove all *their* motions with respect to the center of the Galaxy, which is the only unique reference point shared by all galactic inhabitants. Then it was up to us to offset our receiver frequency to compensate for Earth's motions, to arrive at this unique "galactic" frequency. Because of the uncertainty about the Sun's galactic rotation velocity (we measure it by observing the motions of the stars and gas in our neighborhood), we still had to search a total bandwidth of several hundred KHz. A lot of chart paper was generated during the two years this effort continued, but no recognized signals of intelligent origin were found.

By 1975, a fifty-channel filter bank receiver had been borrowed from the

Green Bank National Radio Astronomy Observatory and software for the already old IBM 1130 computer had been developed by Professor Jerry Ehman (now chairman of the Mathematics Department at Franklin University) and myself, to process all fifty channels continuously. The software was sophisticated, with many internal checks for false alarms and equipment malfunctions. Each of the fifty channels was processed independently, and the computer automatically removed the individual sensitivity variations of each channel. A number of search algorithms were run simultaneously, including searches for both isolated pulses and continuous signals that rose and fell in intensity in just the predicted way (for a continuous, narrow-band signal) as they passed through the antenna beams. The highly processed output data was printed every ten seconds for all fifty channels, and signals the computer thought were "interesting" were also flagged and saved on punched cards for later analyses.

The old IBM computer was built like a battleship and ran without fail for many years. Its operating system could run huge programs in a tiny memory with great efficiency. It was fast, even by today's standards.

Over the years, a few cold hydrogen clouds were found, and huge piles of computer printouts accumulated. There was no magnetic tape drive or equivalent device available, so there was no way to record all the data permanently in computer-readable form. Only the small fraction of data represented by the "interesting" signals was preserved in computer-readable form. Along the way, a small NASA grant was received, which continues today.

Two types of unexplained signals were detected during this search. The first kind is quite rare, with the best example being the "Wow!" signal found in 1977. This name was unintentionally applied from Jerry Ehman's comments in the margin of the computer printout when he noticed the signal. The signal was unmistakably strong, and had all the characteristics of an extraterrestrial signal. It was narrow-band, and matched the antenna pattern exactly, indicating it had to be at least at lunar distance. (A signal from a nearer object would show a wider pattern.) But it was not coming from the direction of the moon, or any planet, or even any particular known star or galaxy. Of course, there are many distant stars and galaxies in the beam of a radio telescope all the time, but that is not significant. A check of man-made satellite data showed that no publicly known Earth satellites were anywhere near the position of the signal source. Furthermore, the frequency of the signal was near the 1420 MHz hydrogen line, where all radio transmissions are prohibited everywhere on and off the

Earth by international agreement. We searched in the direction of the "Wow!" signal hundreds of times after its discovery, and over a wide frequency range. We never found the signal again. It was gone. In fact, while we were receiving it the first time, it turned off as we listened. The radio telescope actually receives two beams from the sky at once (somewhat offset in direction from each other), and subtracts one from the other to cancel out terrestrial radio interference. Objects in the sky are usually received twice with a slight delay, once in each beam. But the "Wow!" signal was received only once, indicating either that it turned off after the first beam received it, or that it turned on after the first beam had passed it.

What was the "Wow!" signal? Probably we will never know. Conceivably it could have been a secret military satellite in solar orbit, transmitting on an illegal frequency (military transmitters often ignore civilian agreements). Its characteristics rule out any terrestrial transmitter, or any near-Earth satellite or reflection from space debris, or equipment malfunction. Perhaps it was a transmission from some other civilization. If so, it seems that they were not trying very hard to attract our attention, since the signal disappeared before we could really find out what it was.

The other kind of unexplained signals we receive are much more numerous. These are narrow-band pulses (lasting less than ten seconds) that go bump in the night. There have been thousands of such signals received, apparently from all over the sky, and never from exactly the same direction more than once. Clearly these signals are not from any single source (intelligent or otherwise), but they are interesting in their own right, and could be some form of previously unknown astrophysical phenomenon. Of course pulsed signals like these could easily be caused by terrestrial radio interference or equipment malfunction. But if that were their source, then they should appear randomly scattered across the sky. The interesting thing is that they do not. They exhibit a zone of avoidance along the galactic plane, and areas of concentration above and below the galactic center, along the galactic north and south polar axes. It is possible that the zones of avoidance and concentration are caused in some complex unknown way by an interaction between the galactic continuum radiation and the automatic gain and correction algorithms in the computer. We simply do not know. A resurvey of a portion of the same area shows roughly the same effect, so the phenomenon appears to be repeatable. We plan to resurvey this area again with all-new equipment in the future.

At one point, there was danger that our telescope would be destroyed. The land under and around it was sold without our knowledge to a real

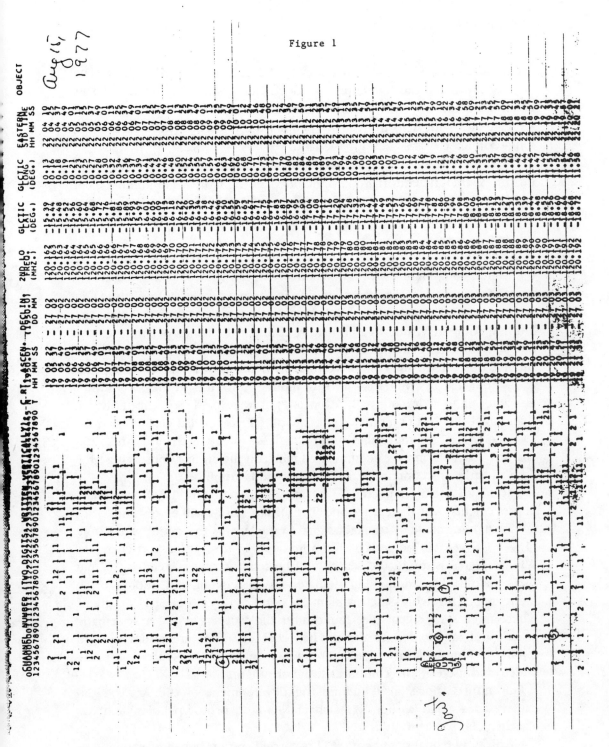

**Figure 1.** The "Wow!" signal. It was recorded on a computer printout of radio noise intensity from 50 frequency channels (digits and letters at left) at varying sky positions. *(Photo: Courtesy Robert Dixon; Ohio State University Radio Observatory.)*

estate developer who wanted to enlarge the neighboring golf course. The developer wanted the telescope torn down and completely removed. This created a furor that was widely reported in the world press. After great struggle and with help from many people, the telescope was saved and a long-term lease was signed for the land.

For several years, we published the first and only SETI magazine, called *Cosmic Search.* Its editorial board included all the worldwide luminaries of SETI. The magazine was a technical and popular success, receiving great praise on all fronts. But it was a financial failure and finally folded after thirteen issues.

In the mid-1980s, a new and more powerful computer was donated by the Digital Equipment Corporation, and we began what we knew would be years of effort to place it into operation in the next generation of the Ohio SETI program. Unfortunately, while this development was proceeding, the old IBM computer came to a premature death at the hands of a mouse. The mouse built a nest at the air intake to the disk drive and cut off the air supply. This caused the disk drive to destroy itself. IBM said the computer was so old that it would cost a lot of money to fix it, and they would not guarantee it to work normally even after it was fixed. So regretfully we abandoned the IBM computer to devote all efforts toward getting the new DEC computer operational. During the years of eight-channel and fifty-channel observations, we accumulated more on-the-air SETI observing time than all earlier or contemporary SETI programs combined.

The new system (expected to be in full operation in 1990) has many improvements over the earlier one. The all-sky search will be resumed with no assumption made as to exact signal frequency, as the entire water hole (1.4–1.7 GHz) will be searched continuously in three thousand channels. When a signal is found the search will be temporarily suspended, so that the signal may be examined immediately in great detail, and studied for an hour or so. This avoids the problem encountered by other SETI programs where interesting signals are found after the fact as part of a systematic search, but are no longer there when reobservations are attempted. An on-line catalog of known Radio Frequency Interference sources is maintained, and used by the computer to ignore them.

Our future efforts also include the development of the Argus radio camera (described earlier by Michael Klein), a prototype of which has already been successfully tested.

Finally, a personal comment. SETI is carried out on behalf of mankind as a whole; the individual people, organizations, or nations involved are not the primary consideration. The *search* is the thing.

**Figure 2.** The OSU radio telescope, with tiltable, flat input radio mirror at right, focusing radio mirror at left, and focal receiver in the small building. *(Photo: Courtesy OSU-RO.)*

# THE HARVARD
# SETI SEARCH

## BY PAUL HOROWITZ AND WILLIAM R. ALSCHULER

**G**iven the overwhelming likelihood of extraterrestrial life, and the plausibility of intelligent life (by which we here mean life that has acquired the technology required to carry out intersteller communication), the central problem in experimental SETI is to predict the mode of communication that the sending civilization may be using to establish contact. During the past few decades there has been general agreement that, among the known possibilities, electromagnetic radiation is the method of choice (though one is certainly free to quarrel even with that choice), given its combination of speed, freedom from interference and absorption, and overall efficiency.

Even if we confine our attention to electromagnetic radiation, however, we are left with too many possibilities—the electromagnetic continuum offers hundreds of gigahertz of bandwidth in the regime called radio waves, topped by hundreds of *tera*hertz of optical bandwidth, and so on up to the wavelengths we conventionally call x-rays and gamma rays. For good reasons the SETI establishment, if we can call it that, has generally held the view, stable now for twenty-five years, that microwave radiation is the optimum method for galactic communication. That conclusion rests upon the observation that at longer wavelengths one must contend with the radio noise pollution of the Galaxy (caused primarily by radiation from electrons spiraling in galactic magnetic fields), and that at shorter wavelengths one must contend with the effects of granularity of detection (or "shot noise") due to the particle nature of the radiation itself. Furthermore, communication through a planetary atmosphere like ours favors the long wavelength portion of this microwave interstellar band, approximately from thirty centimeters to 3 centimeters.

It is worth noting that this view—that centimeter-wave electromagnetic radiation is the galactic communication band of choice, particularly for establishing first contact (as opposed to routine communication, once contact has been made)—is not universally held, and there are thoughtful proponents of infrared laser communication, for example. Furthermore, if one permits the sending civilization to use existing natural radiation (rather than generating the signal from scratch), all bets are off: One could imagine constructing a giant venetian blind shutter, to cause the light seen from our star to wink on and off! In spite of alternative ideas, the advantages of centimeter-wave communication are enormous, hence their popularity: It is a remarkable fact that a pair of three-hundred-meter diameter antennas communicating at a wavelength of 3 centimeters and at a range of one thousand light-years (within which there are roughly a million candidate stars similar to the Sun) would consume only about a dollar's worth of electricity per word transmitted.

A thorough ETI search protocol, then, would consist of a search of the entire sky (or at least the nearest hundred million stars), over the plausible microwave band, looking for radio signals that may be sent in the form of pulses, or carriers (a steady signal—the opposite of a pulse), or some combination ("pulsed carriers"). This is a tall order, and requires resources on a national scale, particularly if one insists on continuous (rather than consecutive) full-sky coverage.

History is littered with overly grand enterprises that tried to plan too much, too early. The same goes particularly for SETI, where we know too little (of the distribution of planetary systems, the prevalence of life, etc.). Furthermore, it isn't exactly easy to mount an effort on a multi-billion-dollar scale, particularly if one is interested in seeing it happen in one's lifetime.

With these considerations in mind, my group at Harvard began a series of modest searches for signals from extraterrestrial intelligence, based on the fundamental assumption that the sending civilization, being more advanced than we, would do their best to make it easy for us to detect their beacon signal. (They *will* be more advanced because: (1) less advanced civilizations do not have the technology to communicate at all—the slice of communicative history behind us is thin; and (2) Drake's equation requires long-lived technological civilizations, if there are to be significant numbers coexisting. Thus, most communicating civilizations will be more advanced; furthermore, the more advanced are by definition capable of more powerful transmissions, further favoring contact.)

And so we seek a guessable signal—one whose characteristics we can deduce without prior contact. In our searches we applied the usual criteria: The method of communication should be efficient (hence microwaves), at a guessable frequency (something like the hydrogen line at 21 centimeters, as originally proposed by Cocconi and Morrison), with guessable modulation (we favor pure carriers, as we shall explain), and guessable allowance for the sender's velocity relative to us.

"Modulation" is an interesting variable—will ET signals be sent as AM or FM, or some sort of sophisticated digital pulse scheme? The choice depends on what the signal is supposed to do.

Information theory teaches us that, in general, the best carrier of information is simultaneously the worst beacon. Thus the interstellar beacon engineer will probably separate the functions of establishing contact (i.e., the beacon) and conveying information (i.e., the communication channel). The beacon is what we seek initially, in the most difficult phase of establishing contact. For this purpose the most likely choices are either a pure carrier or a regular train of pulses. Good arguments can be made for each. We optimized our searches for pure narrow-band carriers because: (1) they are easier to detect, using straightforward Fourier spectrum analysis techniques; (2) they stand out as clearly artificial, being a fraction of a hertz wide in a Universe where the narrowest natural spectral features are at least a kilohertz wide; (3) they are efficient beacons, permitting excellent received signal/ noise ratios through coherent integration; and (4) they permit effective discrimination against terrestrial interference, owing to the particular time-varying Doppler shift signature that is impressed upon a true extraterrestrial signal by the effect of the Earth's rotation.

This matter of carrier bandwidth and Doppler shifts deserves further discussion. A pure (single-frequency) carrier is spread somewhat in frequency after passage through the turbulent charged plasma of the interstellar medium. For distances on the order of one thousand light-years, the spreading is very slight, something like 0.1Hz at the microwave frequencies we are interested in. We can optimize detection of carriers of that width by using a multichannel spectrometer whose channels have a matching bandwidth. We are now faced with a problem, however, because the Earth's rotation causes the received frequency to vary periodically up and down, or "chirp." This is both bad news and good news. The bad news is that we have to work hard and *chirp* our receiver to match the expected signal's chirp; the good news is that interfering terrestrial signals (from the intelligent life rumored to live on Earth) don't have the correct chirp signature, and are thus easily disinguished from the real thing (as well as being weakened enormously by the chirp operation).

Doppler shifts rear their ugly heads in yet another way. Interstellar velocities are quite large—tens to hundreds of kilometers per second, within our galaxy alone—and cause corresponding frequency shifts (which, unlike the "chirp" caused by the Earth's rotation, are unchanging with time) of a few megahertz, at microwave frequencies. We can handle this problem in several ways. One possibility is to build spectrometers of large enough bandwidth to cover all reasonable Doppler shifts; another is to assume the senders have transmitted their signals at a frequency precompensated for their motions relative to a guessable frame of reference. In the first case we need very large numbers of channels—several MHz total bandwidth, divided by channel widths of order 0.1Hz, or several tens of millions of channels; the second case requires many fewer channels; but with the required assumptions as to the intentions of the transmitting civilization, one is, of course, skating on very thin ice.

## NARROW-BAND SEARCHES

We carried out our first search along these lines in 1978 at the Arecibo Observatory's great 305-meter (1000-foot) dish antenna, the world's largest. In that search we looked at 185 Sun-like stars, seeking narrow-band ET carriers with a channel width of 0.015Hz and total bandwidth of 1KHz, centered on the 21 centimeter wavelength of neutral hydrogen. This very narrow total bandwidth encompasses Doppler shifts corresponding to just 0.1 kilometer per second (km/s), which is much smaller than typical relative velocities of nearby stars (approximately 10 km/s); indeed, it is dwarfed also by planetary orbital velocities (Earth's is 30 km/s). In this first search we could not achieve greater bandwidth, given our off-line (no special hardware) spectral computation method. Our solution instead was simply to assume that the sender compensates his transmitted frequency so the signal arrives at the Sun at the true hydrogen frequency. This he does by making spectral observations of our Sun's visible light spectrum, from which he can deduce its velocity along the line of sight. The rest is easy.

This search (*Science,* **201,** 733, 1978) achieved the highest sensitivity ever; it could have detected a 5 kilowatt transmitter, connected to an Arecibo twin aimed at us, at the farthest star examined (eighty light-years away). Radioastronomy deals in weak signals, and this sensitivity corresponds to less than a micro-microwatt ($10^{-12}$ watts) total power falling on

Earth. No signals of extraterrestrial origin were found (bad news), but interference was completely rejected by the narrow-band "chirped receiver" scheme (good news). We were unsatisfied with the restrictive assumptions that we were forced to adopt (precompensated transmitter frequencies), and we felt that the small amount of time available for SETI at busy radio telescopes is not enough. But we were pleased with the interference-free performance, and therefore planned a better search, the so-called Suitcase SETI.

Suitcase SETI (as it has been dubbed) was built at Stanford and NASA Ames Research Center, with support from the latter and the Planetary Society. It consisted of a dedicated spectrometer with 131,072 channels, chirped receiver, and on-line signal recognition and archiving of interesting signals; it achieved twice the bandwidth of the earlier search. We tested it at Arecibo, this time examining 250 favorable stars at another magic frequency, 2841 MHz, the second harmonic of the neutral hydrogen frequency (at the 21 centimeter line itself the radio sky is somewhat noisy, due to galactic hydrogen clouds; furthermore, transmissions at that frequency would interfere with radio astronomy. Both arguments provide plausible reasons to examine the harmonic frequency). We then installed it at the Oak Ridge-Harvard-Smithsonian Observatory, where it operated continuously for over two years, beginning in March 1983.

At Harvard we covered 80 percent of the sky in a third search, an overlapping, nontargeted search called Sentinel. We used a 26-meter (eighty-four-foot) steerable dish antenna, with sensitive amplifiers feeding the modified Suitcase SETI spectrometer. We scanned the sky in "transit" mode—the antenna pointed at the meridian, examining a circle of celestial latitude each day, after which the antenna was moved by one beam width north or south for the next day's circle. At 21-centimeter wavelength, where the beam diameter is one-half degree (and thus each point in the sky spent approximately two minutes in the stationary line-of-sight as the Earth rotated), it took about eight months to cover the whole northern sky. Although the system had one hundred times less sensitivity than Arecibo, it had the advantage of full-time operation and full northern-sky coverage. It is of course impossible to quantify the tradeoff, knowing nothing about the distribution of possible beacons in the sky, or their signal strengths. There may be a very small number of very powerful signals up there, in which case the problem isn't sensitivity, it's persistence.

**Figure 1.** The analysis electronics of "Suitcase SETI," including 64K FFT's at left, and central computer at right. *(Photo: Courtesy Paul Horowitz.)*

**Figure 2.** The META signal processors used to generate real-time radio spectra of incoming signals. *(Photo: Courtesy Paul Horowitz.)*

## WIDER-BAND SEARCHES

After two years' unsuccessful searching with Sentinel, we expanded the system to an 8.4-million-channel spectrometer, covering 400 KHz of total bandwidth (200 times as much as Sentinel) with comparable sensitivity. This new system, called META (Megachannel ExtraTerrestrial Assay), lifts the constraint of assuming the sender has compensated for Doppler shifts relative to our Sun, as demanded by Sentinel and the earlier Arecibo search. Thus META could detect a generalized galactic beacon, not just the Sun-directed transmissions that the earlier searches required.

We built META at Harvard, using twenty thousand integrated circuit chips and a half-million connections (all soldered by hand!). The architecture is a star-connected array of 144 processors, each using a 68000-type central processor, an arithmetic co-processor, and memory. In operation, META cycles through frequencies corresponding to three guessable reference frames, namely the galactic barycenter (center of gravity), the cosmic blackbody rest frame (the frame in which the primordial remnant radiation looks the same in all directions), and the "local standard of rest" (the average motion of stars in our region of the Galaxy). As before, META's chirped receiver weakens local (i.e., worldwide) interference by a factor of about 100, and renders it distinguishable from the expected celestial signature. META has now (fall, 1989) run reliably for four years, during which it has covered the sky several times, at both 21 cm and its second harmonic. Apart from a few interfering signals, nothing has been found. (For technical details on META, see *Icarus* 67, 525-539, 1986.)

## THE FUTURE

SETI should be evolutionary, growing in coverage and sophistication (assuming no signals are found) as technology permits. We are now considering another major expansion of our SETI efforts. Work at Stanford University, sponsored by NASA, has led to a custom-integrated circuit that is efficient for SETI. In particular, one such "DFT chip" can perform more than a thousand 1000-point spectra per second; one chip has the computational power of a first-generation supercomputer (80 million instructions per second). We believe we can build a 100-million channel spectrometer, with 100 MHz total bandwidth, for a cost comparable to what it cost to construct META. Such a system would have 250 times the bandwidth of META, and would permit us to cover the "water hole" (from 1.4 GHz to 1.7 GHz) in three seconds. Such a search strategy would eliminate all require-

**Figure 3.** The 26-meter (84-foot) radio telescope at Harvard. It is being used for a dedicated search for ETI signals. Note the child at bottom for scale. *(Photo: Courtesy Paul Horowitz)*

ments on Doppler shifts, covering as it would a range of frequencies far greater than that caused by any interstellar motions. Of course, even that search makes assumptions—that transmitting civilizations exist, that they are sending microwave carriers in the water hole band, and that signals of sufficient strength arrive at our antenna during the short (approximately two-minute) interval that our beam dwells on the particular point in the sky that is transmitting. One can, of course, build more powerful systems that make fewer assumptions. For example, a phased array of antennas coupled to an array of amplifiers and spectrometers could form a mosaic of the visible sky with a multiplicity of receiving beams, such that a signal of adequate strength, originating *anywhere* in the sky, would be detected immediately. Likewise, larger spectrometers, perhaps coupled to more complex signal-recognition hardware and software, could cover more of the microwave band, sensitive to both carriers *and* pulses. Such ambitious searches are already in the planning stages at NASA, and should be fielded before the end of the century.

(In this work we have had the benefit of the following collaborators: Peter Backus, Dave Brainard, Kok Chen, Joe Caruso, John Forster, Mal Jones, Ivan Linscott, Tap Lum, Brian Matthews, Allen Peterson, Skip Schwartz, Calvin Teague, and Mike Williams. Primary support was provided by the Planetary Society and Steven Spielberg, with small supplementary grants from the Dudley Observatory, the Hofheinz Foundation, and NASA-Ames Research Center.)

# THE U. C. BERKELEY PROGRAM

## BY STUART BOWYER

The central problem in radio searches for signals from an extraterrestrial civilization is the enormity of the search space to be examined. To have any appreciable chance of success, a radio search will have to provide a substantial signal flow in terms of the total search space. Despite the enormity of this potential search task, with few exceptions SETI investigators have had difficulty obtaining substantial amounts of dedicated telscope time. In response to this problem, we developed two automated instruments, SERENDIP I and II (an acronym: Search for Extraterrestrial Radio Emission from Nearby Developed Intelligent Populations), which do not require dedicated radio telescope time but ride piggyback on other work. The SERENDIP systems operate automatically, searching for narrow-band signals. Such a signal would be a consequence of an intentionally easy-to-recognize beacon, or alternatively it could arise as a by-product of other activity, in analogy with the carrier signals of our own radio communications. Our search is carried out in data that is being collected as part of the ongoing astronomical observing program of a radio observatory.

The advantages of this method of data acquisition include very low operation cost per unit observing time, negligible impact on observing schedules, large volume of data collection, and a randomized search strategy that seems desirable since we really know nothing about those whom we hope are attempting to signal us. A piggyback SETI program such as this is not free to choose observing frequencies and sky coordinates. However, in view of the plethora of postulated frequency strategies for interstellar communication and the large number of potential sites for civilizations that have been suggested, this is not necessarily a disadvantage. In fact, some of the frequencies most often observed by radio astronomers have been felt to be logical choices for a transmitting society's beacons, and hence good choices for SETI.

Each SERENDIP system consists of a microprocessor-controlled spectrum analyzer with algorithms for performing simple statistics and digitally recording unusual features in the spectra. In operation, sequential segments of the bandpass being utilized by the astronomical observer at the radio telescope are presented to the spectrum analyzer. The spectrum analyzer measures the distribution of power in the band and the microprocessor notes spectrum peaks on a tape or disk for further analysis. The microprocessor then shifts the band being analyzed to the next sequential band and the process is repeated. Using this procedure we can search through a wide range of frequencies for a possible signal.

The SERENDIP I system was funded at about five hundred dollars per year by the University of California Faculty Grant Fund; it used surplus and loaned equipment, and was quite limited in scope. The primary limitation was the spectrum analyzer, which permitted only one hundred channels of data to be simultaneously sampled. The hardware was developed by a number of graduate students who worked on this project as their master's theses. The system was deployed on the twenty-six-meter (eighty-five-foot) radio telescope at the University of California Hat Creek Radio Observatory in 1981 for a substantial part of a year. Figure 1 shows a plot of frequencies analyzed in one data sample with a power peak (in this case, intentionally introduced) that mimics the effect that an extraterrestrial beacon would have on the data.

Although the SERENDIP I system was limited in sensitivity, we learned a great deal from this effort. Perhaps the most valuable insight was that even in the radio quiet surroundings of the Hat Creek Observatory, we detected a substantial number of narrow-band signals—far too many to allow for any systematic follow-up. After some detective work we were able to determine that many of our signals came from one general direction near the horizon. (This discovery was harder to come by than it might seem, since the coordinates we had recorded were *celestial* coordinates, and only by checking against the time of the observation in juxtaposition with the celestial coordinates did we find this result.) It turned out that twenty miles away behind a series of hills was a small airstrip. We were indeed detecting intelligent life; unfortunately it was terrestrial life, temporarily airborne, in the process of landing. This led us to the realization that any serious SETI effort would have to have some way of dealing with these spurious signals, and our efforts turned to this problem.

At this point, however, we were fortunate in obtaining a NASA grant for our work that, while small in absolute dollar terms, allowed us to develop a vastly improved set of hardware. We dubbed this new system SERENDIP II. The primary improvement made in our system was the addition of a specialized computer unit, called an array processor, to carry out the work previously done by the spectrum analyzer. This unit allowed the simultaneous analysis of about 65,000 separate channels of data. The channel width for each bin (selected on rather general astronomical arguments) was about one Hz (one cycle per second). Upon detection of a possible signal, the civil time, telescope coordinates, bin number, signal power, and other related information were recorded on a floppy disk for later data analysis. The frequency synthesizer in SERENDIP II then stepped up 50,000 Hz in frequency to process the next set of data, allowing a 15,000 Hz overlap between adjacent band samples. (The starting frequency depends on that of the researcher we are working with.)

We made a proposal to NRAO in Green Bank, West Virginia, that SERENDIP II be used with their 92-meter (300-foot) transit telescope (Green Bank was the site of the first modern search for ETI). After receiving extremely enthusiastic appraisals by the NRAO Telescope Allocation Review Panel, we began operation with this telescope in 1987.

The combination of the large aperture of the dish, the exquisite sensitivity of the NRAO electronics, and our powerful SERENDIP II system results in a very sensitive system. The question of the level of sensitivity of any particular search is complex, since it depends upon many factors, including how many frequencies are being searched, how many regions of the celestial sphere are being examined, and the sensitivity of the equipment. It is interesting to note that at a recent international SETI conference the senior scientist making the concluding remarks stated that the SERENDIP II system on the NRAO 92-meter telescope was the most sensitive ongoing SETI program in existence.

During our observations at NRAO, several trillion bins of data have been searched for evidence of a possible extraterrestrial signal. Unfortunately, the many terrestrial and near-space uses of the electromagnetic spectrum cause a variety of interference signatures to appear in the SERENDIP II data records. A typical day's observing produces several thousand events of potential interest—far in excess of the number expected from pure random noise.

The main present thrust of the SERENDIP II project is to explore methods of identifying and rejecting these unwanted signals, while at the same time not reducing the probability of detecting a true extraterrestrial signal. We have reported the details of this work in scientific journals, but some of the techniques are simple to describe. For example, we compare signals detected from one part of the sky with signals from another part of the sky taken shortly thereafter. Many times we find excess power, and hence in principle a potential SETI beacon, at the same frequency from both of these directions. The likelihood that two different technical civilizations would both be signaling us at the same frequency and almost the same time is obviously infinitesimally small. Therefore we can safely conclude these signals are from man-made interference and reject them from further consideration. In all we apply five different techniques to eliminate these man-made signals.

What we find to be truly surprising is that after analyzing our several trillion data samples, we have only about one hundred that stand out from the statistical noise, and that we cannot clearly reject as man-made. This is a reasonable number of events to reexamine; we have, in fact, been granted dedicated telescope time to look at these sources a second time to see if the signals we initially detected are still present.

While we hope that our SERENDIP II system will be successful in finding an extraterrestrial signal (or, in fact, may have already found such a signal!), we are sufficiently realistic to recognize that this is not likely. On the other hand, if we do not search at all, the likelihood of finding such a signal is zero. Even if we do not find the signal we are looking for in the near term (or even in the longer term), the history of scientific discovery shows that progress in any field is usually achieved in painfully small steps. A realistic hope is that we can make a contribution in terms of these painfully small steps.

Over two dozen people, mostly volunteers, have contributed to the SERENDIP system. The core group of personnel, however, are Dan Werthimer, Michael Lampton, Charles Donnelly, Walter Herrick, and myself.

(Note: After this chapter was completed the 92-meter (300-foot) telescope at NRAO collapsed in ruins. It took some time to discover the cause of this failure; it is now known that this was the result of the failure through metal fatigue of an internal structural plate. The tabloid press, however, reported that the telescope was destroyed by alien beings who resented the fact that they were being monitored.

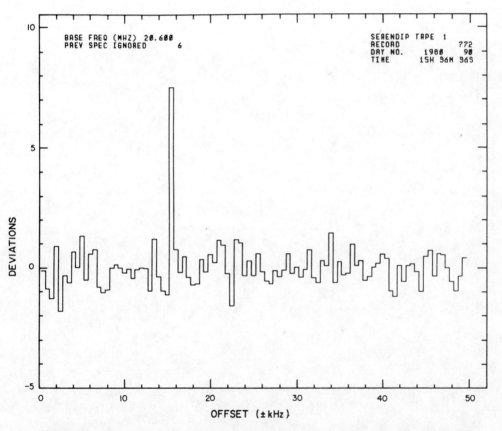

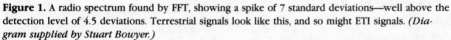

**Figure 1.** A radio spectrum found by FFT, showing a spike of 7 standard deviations—well above the detection level of 4.5 deviations. Terrestrial signals look like this, and so might ETI signals. *(Diagram supplied by Stuart Bowyer.)*

While our friendly aliens may or may not realize they are being monitored, the collapse of this telescope is an obvious problem for our project. We are actively seeking a new telescope and have some confidence that we will be back in operation shortly.)

# CHAPTER 7

# THE POLITICS OF SETI

*There was a day, centuries ago, when scientific research was done by gentlemen of independent financial means. Today scientific research all around the world is funded mostly by government appropriations. The history of SETI, especially in the United States, has been in large part the history of* political *decisions rather than scientific ones.*

*Linda Billings shows in considerable detail how the naivete of scientists combined with the ignorance of politicians for three full decades and more to keep SETI a dream rather than a reality.*

*In a democracy the government reacts to the will of the people—and when the people do not express their will, the politicians do as they please. How do your representatives in government feel about SETI? Do they know how you feel? Unless you make them realize that SETI is important to you, they will be perfectly free to ignore the subject and postpone the inevitable First Contact into the indefinite future.*

# FROM THE OBSERVATORY TO CAPITOL HILL

## BY LINDA BILLINGS

In 1989, thirty years after Cocconi and Morrison published their seminal paper on SETI, planners working on NASA's bedeviled SETI Microwave Observing Project decided to come out of hiding in California and make their way toward Capitol Hill. While supporters of other bigger-ticket, higher-profile NASA projects had been using sometimes-not-too-subtle arm-twisting tactics for years to ensure that they received perhaps a more-than-fair share of the agency's budget from one fiscal year to the next, NASA's SETI team had always been leery of anything that smelled like lobbying. Even as vice-presidents of aerospace mega-corporations camped in Washington for months at a time to protect their space-station contracts, even as other scientists transformed themselves into full-time lobbyists for career-making projects such as Gravity Probe-B, NASA's SETI people kept their noses out of politics.

Not that they didn't pay attention to politics.... Those who had been with NASA's SETI group since the beginning had learned about the stinging winds of politics the hard way, when U.S. Senator William Proxmire, Democrat of Wisconsin, awarded NASA's brand-new SETI program a Golden Fleece award in 1978. And if that wasn't enough to deflate the SETI program, in 1981 Proxmire tacked an amendment onto NASA's Fiscal Year 1982 appropriations legislation prohibiting NASA from spending money on SETI.

By 1989, Proxmire was not only off SETI's tail but retired from Congress. However, NASA's SETI project was by then suffering from a new affliction called Gramm-Rudman-Hollings. The federal budget deficit reduction act had taken a lot of pressure off the members of Congress to decide on the merit of government initiatives case by case and created new headaches for the sponsors of those initiatives. In the Gramm-Rudman era, "last in, first out" became a standard for budget-cutting, and SETI fell victim to it.

Although it will not explain why Washington works the way it does, going back to the birth of SETI will shed some light on how the science of SETI moved from the realm of fiction to legitimacy, giving birth to the idea of a government-sponsored SETI project. The basic concept—the existence of extraterrestrial intelligent life—is as old as the written word. But by 1959, when Cocconi and Morrison published their paper in *Nature,* the Space Age had already begun. Theories of stellar evolution, the formation of planets, and the origin of life had progressed far enough to prompt widespread speculation about the possibility of life beyond the solar system. Thanks to Karl Jansky, Grote Reber, and their successors (who invented radio astronomy in the 1930s), radio astronomy was far enough along to consider using it as the means for detecting artificial signals from elsewhere in the Galaxy. Cocconi and Morrison argued that the only way to respond to both those who believed in the possibility of extraterrestrial life and those who dismissed the idea was to listen for signals.

Publishing a controversial idea in a serious scientific journal such as *Nature* is a sure-fire way to generate a debate and prompt others to expand on the theory. "The science community was somewhat incredulous," Morrison said in characterizing the response to its publication, but the media "were wild for it"; reporters stalked the authors for months. Less than a year later, on April 8, 1960, radio astronomer Frank Drake began the first search for signals from extraterrestrial intelligence with the twenty-six-meter (84.5 foot) radio telescope at the National Radio Astronomy Observatory in Green Bank, West Virginia, prompting another media furor. (Actually, Frank Drake started working on equipment for Ozma in November 1958, before the publication of the Cocconi and Morrison paper. Drake had been interested in extraterrestrial life since childhood. On staff at Green Bank, he calculated that the new telescope might detect signals of the strength Earth was then broadcasting, from sources as far as the nearest stars. He only needed to add some special filters—a few channels—for an extra cost of a few thousand dollars. He proposed this idea to Otto Struve, director of NRAO, the premier optical spectroscopist of the time, who was near retirement. Struve supported the project enthusiastically and gave permission on the spot, but kept it quiet until after the C-M paper was published.) Project Ozma took two months and listened for signals from just two stars—Tau Ceti and Epsilon Eridani—and only in atomic hydrogen's 21-cm spectral lines. The search doesn't seem to be a big deal by today's billions-and-billions standards, but Drake was overwhelmed by the blizzard of media attention that fell upon him. Along with headaches, however, the media blizzard, specifically a *Time* magazine article in 1960, delivered a big boost to Cocconi, Morrison, and Drake.

After reading about Ozma in *Time,* Hewlett-Packard executive Barney Oliver decided to make a side trip to Green Bank on his next visit to Washington. Oliver's interest in SETI went back to World War II and his work with automatic tracking radar. "I was sort of wondering one day," he recalls—what would happen if we sent one-way signals by radar instead of bouncing signals off targets? He concluded, "It would be duck soup to communicate anywhere in the solar system." He immediately considered the possibility of interstellar as well as interplanetary communications.[1] Oliver visited Drake in 1960, and subsequently Drake invited him to the first SETI conference ever organized, to be held in 1961. As it turned out, Oliver and Drake had quite a bit of time to hash over ideas. It was fifteen years before NASA kicked off a SETI program.

Oliver would become a key player in the game of making SETI real, however; after thirty years at Hewlett-Packard in the prominent position of director and ultimately vice-president for research, Oliver retired from H-P in 1983 and went to work for NASA—on the SETI project, of course. In 1989, Oliver and Drake were still worrying about obtaining the funds required to start the search they'd spent their lifetimes planning.

The pioneering SETI conference in 1961, held in Green Bank, was sponsored by the Space Science Board of the National Academy of Sciences. One couldn't ask for a more solid endorsement than a thumbs-up from the SSB, and the conference was an auspicious start for a still very controversial scientific enterprise. Attendees included Cocconi, Morrison, Drake, Barney Oliver, biochemistry expert Melvin Calvin (while at the conference, he received the news that he had been awarded a Nobel Prize), dolphin intelligence expert John Lilly, and Carl Sagan. Soon SETI projects started popping up around the globe. Between 1963 and 1975, the Soviet Academy of Sciences funded several small-scale searches, in both targeted and sky-survey modes. In 1966, Australia's Commonwealth Scientific and Industrial Research Organization funded a modest search, and in the early 1970s small-scale searches were funded in the United States.

---

1. Linda Billings, *"Is Anybody Listening,"* Final Frontier, *Vol. 1, No. 2, 1988.*

While the Apollo program held NASA's attention through the 1960s, scientific discussions continued on the origin of life in the Universe and the possibility of extraterrestrial life. By the late 1960s, at least one NASA scientist decided it was time to get serious about SETI. John Billingham arrived at NASA's Ames Research Center in 1968 as chief of the biotechnology division, following a stint at Johnson Space Center during which, among other things, he invented the liquid-cooled inner garment used in space suits. Ames also had an exobiology division, and Billingham was interested in the work going on there. He picked up the Shklovskii-Sagan book, *Intelligent Life in the Universe,* and liked what he read. In 1969, Hans Mark arrived at Ames as its new director. In making the rounds to meet people, Mark finally hooked up with Billingham, who brought up the subject of SETI: if one were serious about searching for extraterrestrial intelligence, how would one proceed? Billingham was running a summer systems-engineering design fellowship program in conjunction with Stanford University, and he was always on the lookout for interesting study topics. He proposed that NASA undertake a design study of a system for detecting signals from ETI. Mark was intrigued, but he suggested starting with a small study group rather than a full-blown design effort.

Billingham pulled together a small study group in the summer of 1970, and the results looked good. He went back to Mark and recommended proceeding to a bigger summer study, as he'd originally proposed. Barney Oliver was chosen to head the NASA-Stanford project. "We worked like crazy all summer," Billingham recalls. Oliver spent the next year writing up the results. The result was the Project Cyclops report, which proposed a signal-detection system that could start out small and be expanded over time. Some people in the science community reacted with shock; the study appeared to be recommending that NASA undertake a large-scale search for signals of extraterrestrial intelligent origin, building up to one thousand hundred-meter telescopes in an array up to ten miles wide, at a cost of tens of billions of dollars over decades. Astronomers balked at the cost estimates. Project Cyclops was a "what if?" study—in an ideal world, what would be the best way to search for signals?—not intended to produce a practical project plan, Billingham says, but people misconstrued the Cyclops report as a project proposal. SETI scientist Jill Tarter says the Cyclops report had two distinct effects on SETI: it provided a clear scientific justification for conducting a search, and it connected large costs with SETI in the minds of many people. In any case, after the release of the report, Billingham and his colleagues concentrated on educating the Washington establishment and the science community about SETI. Patience, perseverance, a slow and steady process characterized this effort. Ultimately NASA came up with a SETI project plan that would use existing telescopes, develop new signal detection and processing systems, and cost about $100 million over ten years.

Meanwhile, in the early 1960s, Soviet scientists began pursuing the search for signals from extraterrestrial intelligence, spurred by the publication of the Cocconi and Morrison paper and the interest of I. S. Shklovskii, a high-ranking official in the Soviet Academy of Sciences. In 1962, Shklovskii published a book on intelligent life in the universe (later expanded into a joint publication with Carl Sagan, *Intelligent Life in the Universe*). In 1963, Sternberg Astronomical Institute scientists N. S. Kardashev, L. M. Gindilis, and V. I. Slysh formed a SETI group, and in 1964 the first Soviet SETI conference took place at the Byurakan Astrophysical Observatory, sponsored by the Astronomical Council of the Academy of Sciences. Also in 1964, Kardashev conducted the first Soviet search for ETI signals, funded by the academy, and Kardashev, Gindilis, and V. S. Troitskii organized a SETI section of the academy's radio astronomy council. In 1971, the U.S. and Soviet academies of science cosponsored the world's first international symposium on SETI, held in Byurakan. Again, more prestigious sponsors would be hard to find. By 1975, however, SETI fell out of favor at the academy when Shklovskii decided that intelligent life on Earth was probably unique. With his change of position, academy funding for SETI projects dried up, and it was not until after his death in 1983 that official Soviet interest in SETI picked up again.

Back in the States, in 1972 the National Academy of Sciences' survey committee on astronomy and astrophysics for the 1970s came to the conclusion that it was highly likely that intelligent life existed beyond Earth. Also in 1972, another respectable organization gave SETI an endorsement of sorts: At its annual congress, the International Academy of Astronautics held its first SETI review session. Slowly, very slowly, SETI was creeping toward acceptance. The small, tightly knit, and utterly dedicated SETI community was doing everything right, it seemed. But although the idea of searching for signs of extraterrestrial intelligence had gained credence in the science community, it had yet to run the gauntlet of competition for federal funds.

SETI's slow progress at NASA from concept to funded project—twenty years from the time Billingham first proposed a SETI project to the time that NASA headquarters officially initiated it—is typical for a government endeavor. A good idea can percolate for years, especially if it does not have a powerful lobbying force behind it; SETI did not, and still does not. SETI has always been envisioned as a long-term program, stretching out over decades or even centuries, says Philip Morrison, and it's hard for the U.S. government to take on such long-term endeavors, with top officials turning over every four years or less, taking policies and plans with them. Another problem besetting SETI supporters was (and still is) the connection in the public mind with UFOs and bug-eyed monsters; SETI scientists quickly had to become adept at handling flying-saucer fanatics. They also engaged in frequent reexaminations and refinements of their rationale for a search, while constantly testing the waters at NASA headquarters and looking out for potential competitors for funding. (By 1989, for example, SETI was having to compete with two big-ticket space science initiatives that were favorites of NASA's space science and applications chief Lennard Fisk—the Great Observatories and the Back to the Planets campaign.)

In the science community, skepticism about SETI has always been a force to contend with, although it had changed in nature over time and now is certainly manageable. In the 1960s, skepticism was mostly a product of conservatism; older, established scientists weren't wild about the unfamiliar concept, and they were reluctant to advise students to get involved in such endeavors. In 1965, for example, English astronomer Martin Ryle, who chaired the International Astronomical Union's radio astronomy commission, rebuffed a Soviet proposal for international cooperative searches with the comment that SETI would take up too much valuable time at telescopes that would be better devoted to traditional radio astronomy. In the 1970s, skepticism was characterized by cost concerns, fanned by the Cyclops report, and epitomized by Senator Proxmire's Golden Fleece award. Some early SETI advocates changed direction: Ozma II's Ben Zuckerman decided that NASA's high-sensitivity, limited-search approach was not sound; he now favors a larger-scale search for more easily detectable signs of "supercivilizations." By the late 1980s, when NASA had earned the endorsement of the science community for its ten-year $100 million project plan, skeptics in the astronomy community were asking to see more results of the ten years of SETI research already funded by the space agency.

The Cocconi-Morrison argument that the only way to find out if extraterrestrial life exists is to search for signs of it has allowed SETI to endure the slings and arrows of skeptics. No one can prove that life is exclusive to Earth or that life forms capable of developing technology cannot exist anywhere other than on Earth. Therefore no skeptic can state definitely that searching for signs of intelligent extraterrestrial life is a total waste of a scientist's time, just as no SETI believer can state definitively that extraterrestrial life exists. The believers themselves frequently admit that the most comprehensive search imaginable might not turn up any signs, given the infinite Universe that's open to searching. Most skeptics now run along the lines of physics professor James Trefil. In 1981, he coauthored a book *Are We Alone? The Possibility of Extraterrestrial Civilizations*, taking the traditional position of the skeptic by asking, "Where are they?" He wrote: "... the evidence we have at present clearly favors the conclusion that we are alone." Other scientists, of course, had weighed the same evidence and concluded that we are not alone. Trefil worked with the Drake Equation and decided that the odds in favor of the existence of another civilization were virtually zero based on his best estimates of the factors involved. Others examined the factors in the equation and were convinced that it it was time to start listening.

Today, Trefil says that while he still does not believe that a search will be successful, he does not believe that a search should not be made. The SETI community has explained the technical capabilities that are available and what they want to do with them. They've developed a scientific rationale for SETI, and "they've made it good science." Another skeptic, however, has vehemently argued against SETI for years and still shows no signs of giving up. In 1980, mathematician Frank Tipler published a paper in *Physics Today* entitled "Extraterrestrial Beings Do Not Exist." The title says it all: Searching for signals of extraterrestrial intelligent origin is a flat-out of waste of time because intelligent life does not exist beyond the Earth. Tipler, it should be mentioned, is a proponent of the cosmic anthropic principle, which states, roughly, that the Universe exists because humans are here to observe it. The SETI community will never change Tipler's mind, but he will likely remain a chorus of one.

It should be noted that the dedicated people who have kept SETI alive for thirty years have not suffered professionally for being involved in such a controversial enterprise. For example, Barney Oliver spent thirty years as vice-president for research at Hewlett-Packard and won the Presidential Medal of Science in 1986, Carl Sagan is rich and famous, and many give him credit for boosting the stature of his profession by his public speaking and writing. Philip Morrison is still in the midst of a distinguished career in physics and science education. Frank Drake is professor of astronomy and a dean at the University of California at Santa Cruz. In the Soviet Union, Kardashev and Troitskii suffered a bit while SETI was out of favor with Shklovskii and the Academy, but they're back to work on SETI now.

In 1973, SETI researchers developed a project plan based on their feasibility studies of interstellar communication, and they were able to brief then NASA administrator James Fletcher on that plan. The SETI team at Ames presented headquarters with a revised plan in 1974, and Fletcher quickly okayed it; $140,000 was committed to a research program for fiscal year 1975. In January 1975, the team began a series of science workshops, chaired by Philip Morrison, to refine the rationale and weigh the various methods for conducting a search. Workshop participants included scientists from institutions such as Stanford. Harvard, and Cornell Universities, the California Institute of Technology, the University of Michigan, and the Massachusetts Institute of Technology, throwing a lot of weight behind the endeavor.

In 1976, at the sixth and final workshop, Jet Propulsion Laboratory Director Bruce Murray proposed conducting a sky survey—a moderate-sensitivity, wide-ranging search for signals. Ames Research Center scientists had adopted the approach of a targeted search, a high-sensitivity, limited-range option. That year, NASA headquarters promised $775,000 for a SETI research program, to be managed by Ames, in fiscal year 1977. By February 1977, however, it was clear that those funds were not going to materialize. Ames director Hans Mark complained in a letter to John Naugle, chief of space science at NASA headquarters:

"In October 1976 you set an intended agency commitment to SETI of not less than $3/4 million for FY 77. As it now stands we ... will be lucky to see a third of that. Every Program Office has backed out of their [sic] commitment.... I admit that I sometimes can't follow (and even find) the logic behind some program decisions made at headquarters, but this time I am really baffled.... If the Agency's other programs had half the public support and interest and potential that SETI has, your only problem would be how to spend all the funds. I would like to know what your Managers need before they will give active support and assistance rather than ob-struction. I would also like to see the point reached where we can stop responding to the continued requests for program plans in an infinite variety of shapes and colors to meet the weekly funding mark and be able to get down to work."

Mark had a right to be indignant, but what had happened to his fledgling program is not unusual within a government agency. Competing interests constantly jockey for position at the top of an agency's list of priorities, and the stronger one's constituency is at headquarters, the better one's chances are of staying at the top of the list. The SETI program had no constituency at headquarters, so the sharks ate up its money.

For the next couple of years, SETI proponents at Ames and JPL worked on landing funding for a SETI program, each promoting its favored approach. Eventually it became clear that headquarters would be funding only one SETI program; if Ames and JPL were both interested, they would have to figure out how to cooperate. In June 1979, a meeting took place at Ames including Murray, Ames Deputy Director Tom Young, and NASA headquarters officials to work out a way that Ames and JPL could jointly manage a SETI program, and the first joint Ames/JPL SETI program meeting was held at Ames in July.

Despite its start-up troubles, which were not out of the ordinary, SETI was certainly beginning to look respectable. Through all of the discussions and debates of the 1960s and 1970s, the scientific basis for proposing to listen for radio signals from extraterrestrial intelligence solidified: Planets appeared to be a natural result of the formation of stars, astronomers had detected organic compounds in interstellar space, the laws of physics and chemistry indicated that under the right set of conditions life might begin anywhere. The well-known writer Norman Cousins concluded: "It is almost unscientific to think that life does not exist elsewhere in the Universe. Nature shuns one of a kind."[2]

But in 1978, disaster struck. Senator William Proxmire awarded SETI a Golden Fleece award, his way of publicizing cases of government waste. In his opinion, SETI research was a waste of taxpayers' money; there was nobody out there, Earth inhabitants were alone in the void. NASA attempted to rally behind its program after the fact, and later that year the House Science and Technology Committee, which was responsible for authorizing new programs at NASA, held two days of hearings on SETI, stressing that it was a part of the study of the bigger question: What is the nature and distribution of life in the Universe? SETI suffered but did not die from the blow of the Golden Fleece award; NASA spent $300,000 in 1979 and $500,000 in 1980 on continued SETI studies, and the agency began drafting a SETI program plan.

In 1980 NASA formed a SETI Science Working Group, expanding the advocacy group for SETI and further legitimizing the program. Members included Frank Drake, MIT physics professor Bernard Burke, and astrono-

2. Norman Cousins, Why Man Explores, *Washington, D.C., NASA EP-125, 1976.*

mer Benjamin Zuckerman. Tim Mutch, chief of space science at NASA headquarters and favorably inclined toward SETI, produced guidelines for a NASA Life in the Universe program that would include SETI under a ten-year plan, with the program to begin in fiscal year 1982. The program seemed on the verge of establishment. But then the SETI people suffered a setback that can only be described as a cosmic fluke: In November, Mutch died in a mountain-climbing accident. The SETI program plan was still sitting on his desk, unsigned. In 1981, NASA headquarters cut the ten-year plan in half.

By fiscal year 1981, SETI was allotted $1 million to begin work on prototype hardware and software—multichannel spectrum analyzers that would permit large-scale, real-time time signal detection and processing, and programs that would manage the signal-processing operation, characterize signals, and weed out known natural and artificial sources of radio signals. A few months into the fiscal year, a National Academy of Sciences survey committee headed by George Field and appointed to recommend astronomy and astrophysics programs for the 1980s aired its findings, which included a strong endorsement for SETI. The Field Committee endorsement was valuable; Congress and the administration depended on the Academy's survey committees to provide guidance. By this time, the technical basis for SETI was bolstered by the rapid advancement of the digital data processing technology that would enhance such a search, and the Field Committee made note of that important fact:

"Our interest in the tiny fraction of the matter in the solar system that condensed into planets is heightened by the fact that life has developed on at least one of them. Have condensations to planets and the origin of life occurred elsewhere as well? And has that life evolved into communicative intelligence, with which we human beings might be able to enter a conversation about life in the Universe?

"These questions reach far beyond astronomy, and even beyond science as we currently think of it. Yet astronomers, who are in a sense commissioned by the public to keep an eye on the Universe, feel bound to ask them and to point out how we might begin to try to answer them."[3]

In 1982, the International Astronomical Union formed a bioastronomy committee, Commission 51, to monitor SETI developments and the search

---

3. *G. Field ed.,* Astronomy and Astrophysics for the 1990's. Volume I: Report of the Astronomy Survey Committee, *National Academy Press, 1982.*

for extrasolar planets and the study of the origin of life in the Universe. (It is now the second largest commission of the IAU.) Commission 51, headed by Michael Papagiannis, held its first international symposium in 1984. This prominent group would prove to be a reliable advocate for NASA's SETI project.

Unbeknownst to NASA's SETI people, however, Senator Proxmire still had a jaundiced eye fixed on SETI. In July 1981, the senator tacked an amendment onto NASA's fiscal year 1982 appropriations legislation prohibiting spending on SETI, and the amended bill was signed into law. The Golden Fleece was a slap in the face, but this move was strangulation. While officials at NASA headquarters and the SETI team in California pondered how to keep the program alive, astronomer Carl Sagan, president of a newly formed space exploration advocacy group called The Planetary Society, quickly scheduled a meeting with Senator Proxmire. Proxmire respected Sagan's professional reputation and had dealt with him before, regarding the prospects for nuclear war. In this meeting, Proxmire reiterated his belief that life did not exist beyond Earth, and Sagan responded with a discourse on the Drake Equation.

Proxmire soaked up what Sagan had to say about the number of galaxies in the Universe, the number of stars in a galaxy, the distance between stars, the evolution of life, and the lifetime of a technological civilization. "After talking with Sagan," Proxmire explained in 1989, "I took another more careful look" and decided that the rationale behind SETI was worth a second consideration. He sent a list of questions to NASA regarding the science and technology involved in SETI and the prospects for international cooperation in a search. According to the senator, the combination of Sagan's efforts and NASA's explanations finally convinced him that there was enough sense in the SETI program to justify the small amount of funding NASA was seeking. Proxmire still asserts, however, that "there's absolutely no evidence whatsoever" that life exists elsewhere in the Universe. By the end of September 1981, just a week before the Proxmire amendment would go into effect, NASA headquarters informed the SETI team in California that Proxmire had decided he would not resist another request for SETI funding.

The SETI community was still not absolutely certain that its troubles were over, however. In October 1982, a petition endorsing SETI was published in *Science* magazine.[4] Led by Carl Sagan, seventy-one science

---

4. Science, *CCXVIII, no. 4571, 1982.*

luminaries, including seven Nobel laureates, had signed "Extraterrestrial Intelligence: An International Petition." Freeman Dyson, Stephen Jay Gould, Stephen Hawking, Fred Hoyle, Linus Pauling, Roald Sagdeev, Lewis Thomas, and others, asserted: "We are unanimous in our conviction that the only significant test of the existence of extraterrestrial intelligence is an experimental one.... We believe ... a coordinated search program is well justified on its scientific merits." Although this petition by no means went unnoticed, the combination of Sagan's arguments and NASA's explanations—collectively known as "advocacy"—had already convinced Proxmire to back off from attacking SETI. Congress appropriated $1.6 million for SETI in fiscal year 1983, the first year of a five year research and development program. In 1983, the SETI Science Working Group published its final report, stating once again: "... it is timely and promising to conduct a search for other civilizations by testing for the presence of short-wavelength radio transmissions from them."

How did the SETI initiative get into such an awful mess on Capitol Hill? For one thing, it appears that NASA had not explained the SETI program on Capitol Hill before asking Congress to fund it. Officials at NASA headquarters are often wary of "selling" to Congress. NASA officials still complain that they can't air plans for new programs until they are funded, and they can't get them funded unless they have plans. NASA's congressional relations staff follows the lead of top-level officials at headquarters, tending not to offer information until a member of Congress asks for it. The SETI people at the California field centers were wary of doing anything that might be perceived as lobbying, so they stayed home and kept quiet. The result was that the SETI program was almost stillborn.

Despite the troubles with Congress, by the late 1980s prospects appeared to be brightening for NASA's SETI group. Astronomers such as David Latham of the Harvard/Smithsonian Center for Astrophysics and Bruce Campbell of the University of Victoria in Canada and William Forest of the University of Rochester reported detecting optical evidence of extrasolar planets, lending more credence to the idea of extraterrestrial life and boosting public interest in the subject. A privately funded search for extraterrestrial intelligence called Project META, conducted by Harvard University physics professor Paul Horowitz and funded by the Planetary Society and filmmaker Steven Spielberg, started up, and no one poked fun at it.

In the fall of 1987, Barney Oliver, now a member of the SETI team at Ames, was called to Washington to brief White House Chief of Staff Howard

Baker on his program. This meeting came about thanks to White House Science Adviser William Graham: As a graduate student at Stanford University in the 1960s, Graham had heard Oliver, a Stanford alumnus, lecture on SETI, and he'd been interested ever since. When Graham heard that Baker had an interest in SETI, he knew who to call. By early 1989 Oliver had been back to the White House, at Graham's invitation, with other members of the SETI team, to brief a group of administration officials on SETI. While it's certain that none of these White House meetings hurt the SETI cause, it's not at all clear that they were of any particular help. For all of his public endorsements and policy pronouncements, President Reagan was not committed to the space program, let alone SETI, and his space-policy-making group, an arm of the National Security Council, never put SETI on its agenda. In fact it had been run quietly by the NSC for most of Reagan's term, but has recently revived as the National Space Council and provided policy support to President Bush's July 1989 space goals.

In October 1987, NASA Administrator James Fletcher made a keynote speech to the International Astronautical Federation in England that electrified much of his audience. The speech was radical, maybe even visionary, for NASA: "I'd like to look ahead—thirty years into the future—at two priority subjects that NASA has thought a great deal about.... The first is philosophically profound. It is the necessity to put real effort into the search for extraterrestrial intelligence...." Speculating on what kind of speech a NASA administrator might give in 2017, he predicted that what we will have learned about the Universe will point more and more to the inevitability of extraterrestrial life. More than six full pages of his speech were devoted to SETI. Members of the SETI community who heard the speech were dumbfounded, and given their long history of neglect at NASA headquarters, they could not help but wonder whether the speech was an indication that Fletcher would go home and do something about funding a SETI project. As it turned out, Fletcher did nothing to protect the 1989 budget request for SETI.

In March 1988 Fletcher received a letter congratulating him on including an augmentation of funds for a SETI project in the FY 89 budget. Signatories were Carl Sagan, Freeman Dyson, Stephen Jay Gould, Hans Mark, Philip Morrison, and Thomas Paine (former administrator of NASA and chair of the National Commission on Space). "If you need support at any time in defending SETI before Congress, we would be happy to do so," they said in conclusion. The letter turned out to be premature, though; late that summer the appropriations committees cut NASA's budget in a way that ensured SETI would not receive the money it needed.

From 1984 through 1987, NASA's SETI program had received $1.5 million to $2 million a year for research and development. The R&D results pointed to the same conclusion SETI scientists had been coming to for years. To quote SETI scientist Jill Tarter, "It's time to stop talking and start listening." In 1986, a proposal for a full-fledged SETI microwave observing project passed a critical review at NASA headquarters, and by 1987 the SETI team had drafted a project initiation agreement. By the spring of 1988, the agreement bore all the necessary signatures, but there was more trouble in store on Capitol Hill.

In early 1988, NASA asked Congress for $6 million to fund the first year of preparations for a full-fledged SETI microwave observing project in Fiscal Year 1989. The $4 million-plus augmentation to the SETI budget flashed like a red flag in front of the eyes of deficit-crazed legislators. NASA headquarters had not bothered to keep congressional staffers posted on the status of the project, so the decision was made to let SETI wait another year. Little did the people on Capitol Hill know about how long the SETI team had waited, and waited, and waited, for headquarters to scrape up a budget for them. The 1989 SETI budget turned out to be the same as 1988's.

No members of Congress had decided to shrink the program. But although congressional staffers reported that SETI had not been singled out for budget-cutting, the SETI team took the hit personally; they could not start building hardware without money, so they would have to wait until next year for another shot at the funds needed to shift out of the research phase and into full-scale development. They kept working hard, since their ranks kept dwindling due to limited funds. And plotting commenced to ensure success next year.

Meanwhile, as SETI fared badly in Congress and at NASA headquarters, it was thriving in the public eye. In 1988, the National Air and Space Museum began planning for a major space exhibit to open in 1992, detailing the next five hundred years of space exploration and exploring SETI as a major theme. The museum had also invested a quarter-million dollars in a new planetarium show on SETI, to open in 1989. At the same time, the Pittsburgh public television WQED began researching a major series to be aired in 1992 called *Space Age.* Again, SETI would be a major theme. Newspapers, magazines, and other media outlets continued to carry articles on SETI. In 1989, NASA's 1990 budget request went to the Hill, once again including the augmentation of funds required to gear up the SETI project. The project's data for start of observations was October 1992. By now the SETI team had grown tougher and wiser, and when SETI still appeared to be threatened, by Gramm-Rudman-Hollings budget-cutting fever, they decided not to watch from the wings this time. Instead, they trekked their way to Washington to make their case. The group included Michael Klein, who heads up JPL's SETI team, who made several important visits to Washington D.C. on behalf of SETI.

At this point, Ames Research Center and the Jet Propulsion Laboratory were co-managing the SETI project; Ames was planning the targeted search and JPL the all-sky survey. The SETI Institute, a nonprofit organization handling SETI and other research projects under cooperative agreement with Ames, employed several key members of the SETI team. The Planetary Society in Pasadena was supportive of NASA's project, but it was putting its money behind other search projects (Project META in Massachusetts and an Argentinian initiative). Thanks to SETI's budget troubles, several key members of the SETI team at Ames were no longer on NASA's payroll, although they were still on the team. Their status as private-sector employees left them free to contact politicians without waiting for invitations. Tom Pierson, director of the SETI Institute, and Jill Tarter, chief scientist for the SETI project and an employee of the SETI Institute, plotted a campaign to ensure that funding for the "operational" phase of the SETI Microwave Observing Project materialized in 1990. Until they began their public education campaign, Tarter and Pierson didn't quite believe that their project had no "enemies" in Congress. Scars caused by the Proxmire-Golden Fleece assault were taking a long time to fade. Nonetheless, they came up with a simple message to take to the Hill which turned out to be very effective—their project had been fully authorized for 1989, but appropriations had not been provided to match this authorization. Given the go-ahead from Congress, the project was nevertheless paralyzed by lack of funds and would remain so unless Congress appropriated the $6.8 million requested for SETI in 1990.

In January 1989 Pierson and Tarter made their first assault on the Hill, meeting with a handful of staffers. In February Pierson and Tarter returned to Washington for another round of congressional meetings. They also consulted with the National Science Teachers Association and the University of California System's Washington office (Tarter was also affiliated with U.C.-Berkeley). And they found a friend in Representative Tom Campbell, a freshman Republican elected from the Twelfth District of California, which includes the Ames Research Center, home of the SETI project. Representative Norman Mineta, a seasoned Democratic representative from the Thirteenth District of California whose constituents included many employees of Ames, turned out to be a friend, too. In April, Pierson had a chance encounter with Mineta at Dulles International Airport as both were waiting for flights from Washington back home. In his usual low-key but quick-witted fashion, he made the most of the moment. He had just viewed the National Air and Space Museum's new planetarium show on SETI. Mineta was a member of the board of advisors of the Smithsonian Institution, which funds the museum. Mineta was impressed by Pierson's report on the amount of money invested in the show ( $250,000 ), the size of the audience that would see it (hundreds of thousands, at least), and the rave review that it had earned in the Washington *Post.* Pierson got on his plane knowing that he could call on Mineta for help.

Overall, Pierson and Tarter found their audience on Capitol Hill uninformed but receptive to their message. The perception that some staffers had no idea what SETI was about or where it stood at NASA caused Tarter to write a blistering letter to the editor of a Washington weekly newspaper, when it ran a story about SETI on April 14, 1989, paired with another feature about a UFO nut-case. The letter was published a few weeks later: "Those of us who work on the NASA SETI Microwave Project are well aware of the effort that has been expended over the past two decades to distinguish the science of SETI from the pseudo-science and outright fraud that characterizes the UFO community. Your editorial manipulations ignored and attempted to obscure this important distinction. In any other city this might be harmless, and I would just accuse you of sensationalism in an attempt to sell more advertising space, but in Washington such action is irresponsible as it can threaten federal funding. No elected official or staff person who has had the opportunity to be briefed on the NASA SETI Microwave Observing Project ... could be confused by your merging of SETI and UFO stories. Unfortunately, many members of Congress and their staff members have not had the opportunity for a SETI briefing, and being short on time, could well gain the wrong impression...."

At the end of 1989, the problems facing NASA's SETI project were uncertainty regarding support at NASA headquarters, uncertainty over how little money it could get by on compared with what was requested for 1990, and a continuing need for public education regarding what SETI is all about. After investing $10 million in a ten-year SETI research program, what's available from NASA are designs and early prototypes for signal detection and processing systems, products that are not readily visible to the astronomy community. Hence, some astronomers still question the value of the program; where's the data?

Officials at NASA headquarters, always interested in publicizing the "spin-off" benefits of their programs, have not promoted the broad variety of applications for SETI technology, although recently SETI officials at the Ames Research Center have been exploring the use of SETI technology for air traffic control systems. Although no consensus has developed on this point, some radio astronomers already anticipate that radio frequency interference surveys, automated observatory control programs, and other components of the SETI Microwave Observing Project will be of great value to the radio astronomy community. Meanwhile, a privately financed SETI project begun in 1985, the Megachannel Extraterrestrial Assay (META), is demonstrating the real-time high-speed parallel-processing concept behind NASA's SETI project. META is run by Harvard University professor Paul Horowitz, who got his start in SETI as a National Research Council Fellow at the Ames Research Center; META searches 8.4 million channels simultaneously, an approach similar to the one that NASA is taking.

Field tests of prototype SETI equipment, begun in 1989, may help to turn some skeptics around, says Jill Tarter. The public is still very interested in SETI, and SETI still draws a lot of attention in the press. The National Academy of Sciences' new survey committee on astronomy and astrophysics for the 1990s is interested in the status of NASA's SETI project. And as the project moves into its development phase, John Billingham, who has called himself "the SETI obstetrician," is starting to worry about funding for continued research into new search strategies and other innovative SETI concepts. In the SETI community, sentiments haven't changed much over thirty years: "Let's get on with it."

# CHAPTER 8

# FIRST CONTACT— SEIZING THE MOMENT

*What happens after we make contact?*

*The world will change once we make unequivocal contact with another intelligent species. Our individual lives will change. Our societies will change. Perhaps not immediately, perhaps only imperceptibly at first. But the world will never be the same afterward. The change will be as enormous, and as permanent, as the results of the contacts between Europe and America that started five hundred years ago. Michael Michaud takes us on a tour of some of the possibilities.*

*"Answer, Please Answer" is also an examination of what happens after we make contact. Written in 1962, when the Cold War confrontation between East and West was at its most frightening (the Cuban Missile Crisis occurred within weeks after the story was published), this science-fiction story deals with the motivation behind SETI; the reason why a race of intelligent aliens might be desperately trying to reach another intelligent species.*

# A UNIQUE MOMENT IN HUMAN HISTORY

## BY MICHAEL MICHAUD

Consciously and unconsciously, we are making contact with extraterrestrial civilizations more likely. The evolution of our technological civilization is making Earth electromagnetically noisier, with radio signals, television carrier waves, and radar pulses radiating outward into the Galaxy. Though we may not intend to call the attention of other civilizations to ourselves, we have been doing so for most of this century. We are making it more likely that other intelligences—if there are any—will find *us*.

At the same time, we have embarked on our own searches for life and intelligence beyond Earth, first with optical telescopes, then with planetary probes and radio observatories. So far, we have failed to find convincing evidence of another civilization. But, by extending the sensitivity and duration of our searches, we are making it more likely that we will find *them*.

Because we search for others, we tend to assume that they search for us. If we scour the skies with our instruments, send automated probes to other planets in this solar system, and imagine sending such probes to the planets of other stars, we assume that others are doing the same.

Yet this search for others may be an episodic cultural phenomenon in our own civilization, dependent on the values and perceptions of the time. The idea of a plurality of inhabited worlds has had its ups and downs throughout recorded human history; sometimes it was widely believed, and at other times it was widely rejected. This implies that other civilizations, if they ever start such a search, may not give it continuing attention over the millennia, particularly in the absence of a positive result. Thus the detection of another civilization, by us or by them, may not be the result of a thoughtfully planned search conducted by astronomers sympathetic to it; it may come as a *surprise*. It may be the unintended by-product of other activities, such as astronomy, planetary or interstellar exploration, or the gathering of military intelligence. While there are strong arguments for radio as the preferred method of search, we should not exclude the possibility of other scenarios, such as finding an artifact of another civilization in our own solar system, or spotting the exhaust trail of an interstellar spacecraft. Those too would be forms of detection.

Given the youth of our own technological civilization, probability suggests that alien civilizations capable of detecting or communicating with us would be older than ours and technologically superior. This suggests that they are likely to find us before we find them.

## THE CONSEQUENCES OF CONTACT

What happens if we do detect others, or meet them face to face? Because of the probable technological superiority of the alien civilization, there is a presumption that the relationship will be an unequal one, implying a submissive reaction on our part. But the consequences of contact depend heavily on the circumstances of the detection, and the state of our own civilization at the time.

At one extreme is the classic radio astronomy scenario, in which our radio astronomers detect a faint signal that is the product of another intelligence. After lengthy efforts, a message is deciphered, and the wisdom of a superior civilization is revealed to us. The remoteness of the aliens, perhaps hundreds or thousands of light-years away, implies that they will be no threat to us, and that an exchange of information may be the major outcome of contact.

At the other extreme is the direct contact scenario, in which an alien spacecraft touches down on Earth, and we encounter extraterrestrials face to face. As envisioned in science fiction, the aliens could be as benevolent as the cute alien botanist E.T., or as malevolent as the marauding invaders depicted in the paranoia-charged atmosphere of the early 1950s.

In our thinking about aliens, we reveal our emotional selves—our predilections, our preferences. We are variously hopeful, naive, hostile, intolerant; we display idealism, wishful thinking, insecurity, fear, defeatism, even self-loathing. At one extreme, we think of aliens as altruistic teachers who will show us the road to survival, wisdom, and prosperity, or God-like figures who will raise humanity from its fallen condition. At the other, we see the aliens as implacable, grotesque conquerors whose miraculous but malevolently applied technology can only be overcome by simpler virtues.

These images are determined largely by our cultures, and by the circumstances of the time. Consider how American film and television portrayals of extraterrestrials changed from the weird and horrible invader of the 1950s (*The Thing, Invasion of the Body Snatchers*) to the benign aliens of the 1970s (*Close Encounters of the Third Kind*), and then back to the repugnant aggressor of the 1980s television series *V* and *War of the Worlds.* We carry these images around in our heads, and they will influence the way we react to contact. (Of course, in other cultures people may carry other images.)

Our emotional and intellectual predispositions could be reinforced strongly by contact. Those humans who suffer deeply from guilt, who think that our species is uniquely evil, may fear retribution, a chastising of humanity; some may even welcome it. Those who despair at humanity's lack of wisdom, or who are frustrated by important unanswered questions, may see in the aliens a long-desired source of guidance and solutions, a living, law-giving deus ex machina. Those who perceive contact in the context of the more brutal episodes of human history may fear attack, invasion, or enslavement. We are likely to attribute motives to the aliens before we have real evidence.

Contact almost certainly would cause many more humans to attribute events on Earth to alien intervention (some already see this in the UFO phenomenon). There might be an upsurge in conspiracy theories, witch-hunting, and UFO sightings. But many of us would simply be excited by this new outside stimulus, with its suggestion of a break with conventionality and of new prospects for the future. Contact could be shared adventure for a species that badly needs one.

## ANTHROPOCENTRISM GOOD-BYE

The most profound message from the aliens may never be spoken: We are not alone or unique. Contact would tell us that life and intelligence have evolved elsewhere in the Universe, and that they may be common by-products of cosmic evolution. Contact would tend to confirm the theory that life evolves chemically from inanimate matter, through universal processes, implying that there are other alien civilizations in addition to the one we had detected. We might see ourselves as just one example of biocosmic processes, one facet of the Universe becoming aware of itself. We would undergo a revolution in the way that we conceive our own position in the Universe; any remaining pretense of centrality or a special role, any belief that we are a chosen species would be dashed forever, completing the process begun by Copernicus four centuries ago.

The revelation that we are not the most technologically advanced intelligent species could lead to a humbling deflation of our sense of self-importance. We might reclassify ourselves to a lower level of ability and worth. This leveling of our pretensions, this anti-hubris, could be intensified if we were confronted with alien technology beyond our understanding. (Arthur C. Clarke has observed that any sufficiently advanced technology would be indistinguishable from magic.) We could feel even more deflated if the aliens, after contact, showed no interest in talking to us.

Contact also could be immensely broadening and deprovincializing. It would be a quantum jump in our awareness of things outside ourselves. It would change our criteria of what matters. We would have to think in larger frames of reference. Continuing communication with an ancient civilization would strengthen our sense of our own genetic and historical continuity, and could encourage us to take on longer-scale projects than we do now. Awareness of extraterrestrials would help to establish a new cosmic context for humankind; we would leave the era of Earth history and enter an era of cosmic history. By implying a cosmic future, contact might suggest a more hopeful view of the Universe and our fate, one less alienating than the cynical, materialistic, and limiting visions of the present.

Contact would remind us, as nothing else could, of our identity as a species. We would see the common nature of human beings defined by contrast with the aliens; the racial, religious, linguistic, and cultural differences among humans would seem minor by comparison. This could have a considerable unifying effect on humanity, easing tensions and encouraging cooperation within our species. But this new unity could be based as much on shared fear as on a sense of human brotherhood. If direct contact occurred, it could lead to a new racism, directed against the aliens.

Contact would give us the satisfaction of making others aware of our existence. If we detected extrasolar aliens, we would be strongly tempted to send a signal immediately to announce our presence, tell the aliens about ourselves, and begin spreading our own culture and values. But we have many causes for embarrassment about human civilization and behavior, and we might be tempted to disguise our problems and engage in posturing, inflating our stature and conveying an image of perfection. The aliens might not be above doing this themselves.

Contact also would be very reassuring to a species as doubtful about its future as we are. It would tell us that life and intelligence had survived and prospered elsewhere, even after acquiring powerful technologies. If the alien civilization were superior to ours, contact would suggest that intelligence is not an evolutionary dead end, and that the present state of human development is not final. More than any other event, contact could motivate us to transcend our present condition.

Contact would end the isolation of our species from other minds, giving us a new perspective on intelligence and on ourselves. At last we would encounter other beings who also worry about their survival, who feel the pain and joy of awareness, and who seek answers to many of the questions we ask about the purpose and destiny of intelligent life. We might enter a community of intelligence, gaining access to new knowledge and sensibilities, participating in a vast commerce of ideas among disparate minds. And we might join together with other civilizations in a mutual effort to assure the long-term survival of intelligence in the Universe.

## THE KNOWLEDGE REVOLUTION

Contact could bring a knowledge revolution. Simply detecting aliens would bring us new knowledge about the evolution of life and intelligence, especially if we could identify the characteristics of their home star and planetary system. Even undecipherable signals could tell us much about their technology and their command of energy. Radio communication could allow exterrestrials to transmit vast quantities of information deliberately. Philip Morrison has suggested that aliens might send us a volume of information greater than that transmitted to medieval Europe from the ancient Greeks, stimulating a new and even greater Renaissance. By entering a communications net, we might receive maps of the Galaxy, and elaborate descriptions of the physical Universe and how it works. We might learn the histories of civilizations stretching far back into the galactic past, and become aware of alternative cultures, arts, social and economic systems, and forms of political organization. Deliberately or by implication, the aliens might tell us how they had survived. It is intriguing to consider how much we could contribute to the other side of the dialogue.

Alien knowledge, integrated with our own, could generate a dramatic forward leap in our sciences and our other academic disciplines. For the first time, we could compare our information and our perceptions with those of other minds in different environments, illuminating voids in our own knowledge and suggesting new generalizations. This almost certainly would lead to new syntheses, a boom in interdisciplinary studies as we perceived new linkages, and new branches of science. Dealing with this influx of new knowledge could force us into mind-stretching responses. Our curiosity would be stimulated by finding out how much we had not known. Contact also could reveal areas of shared knowledge, supporting our own conclusions; this might include religious concepts such as creation or a Supreme Being.

But we should beware of excessive optimism about this exchange of information; communication with an alien civilization may not be easy. No matter what we *wish* to believe, aliens, by definition, will be very different. While they may share some of our perceptions of physical reality and some of our evolutionary experiences, their evolutions would differ from ours in many ways, and we might share little in philosophy and culture. There could be serious problems of mutual unintelligibility, or misunderstandings caused by different ways of perceiving reality and by different cultural frames of reference. We might find that our own concepts of language, including mathematics, are narrow and idiosyncratic.

We also should not assume that the aliens will want to tell us everything. Transmitting the species data bank might not be the aliens' first priority. They might want to know first our capabilities and our intentions to assure themselves that their security would not be threatened. There might be things they would not want to tell us, such as how to achieve interstellar flight or how to create more powerful weapons.

Receiving knowledge much more advanced than our own, and the solutions to problems we have struggled with for years, could break the intellectual morale of some scientists and other scholars, and undermine support for some forms of research. Instead, we might simply wait for alien answers, and translate them into our terms. Humans concerned about their personal and institutional interests might resist the dissemination of some alien information, or seek to brand it as dangerous, immoral, or subversive.

Receiving, interpreting, and disseminating information from extraterrestrials could be a major enterprise for humanity, almost certainly requiring new institutions. Since control over this information could bring great power and status, there would be a strong temptation to monopolize the channel and to limit access by others. Individual nations or groups might attempt to conduct separate dialogues with the aliens to exploit contact for

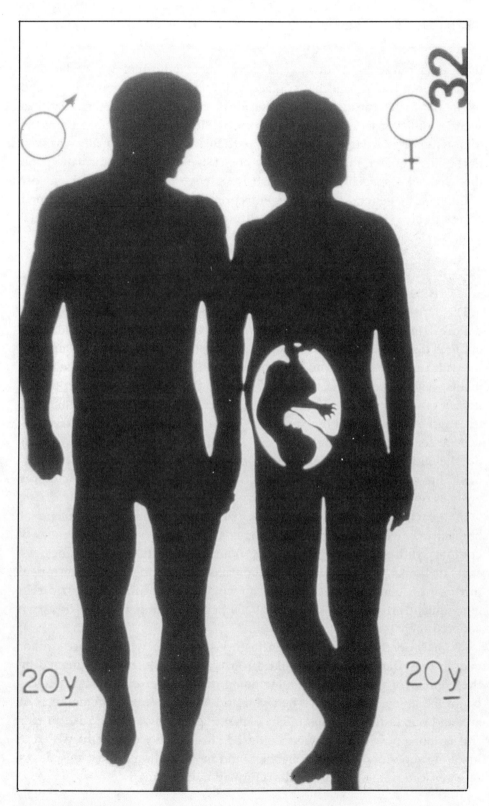

**Figure 1.** An image from the record sent along with the Voyager spacecraft, now on its way into interstellar space. *(Photo: Courtesy The Astronomica  ociety of the Pacific.)*

their own purposes. Political and governmental leaders would be concerned about the impact that contact could have on their populations, and might try to let through only those ideas they considered safe. National security policy-makers might argue for classification of the contact and the information received. Some scholars, particularly those personally involved in the first contact, might be equally possessive about the information and the channel, especially if they distrusted governments and held a low opinion of the general population. Entrepreneurs might compete to get first access to alien ideas and to monopolize or patent those with commercial value.

## THE FATAL IMPACT

The more intense forms of contact could have a fatal impact on our culture. Human history is littered with examples of cultural shock, of cultures that were destroyed or absorbed by other civilizations. An encounter with superior aliens could disorient our thinking, diminish our achievements, and shake our confidence. Even if the aliens meant well, their impact on us could amount to cultural imperialism; the missionary mentality may not be uniquely human. If the aliens were experienced in contacts with lesser civilizations and were concerned about the damage they might do, they might seek to reduce the shock of contact, or even avoid continuing it. But our own record in dealings between unequally powerful cultures gives us no reason for optimism.

In the cultural sense, contact could be the beginning of the end of humanity as we have known it. Contact's stimulus could produce a new cultural synthesis, leading to a new human civilization. Over time, our separate human culture might fade and vanish, becoming a quaint historical memory as it merged with a superior culture. Our anthropocentric religions might crumble, as superior aliens became our new gods, or as we adopted their religious concepts.

We have learned from our own history that a receiving culture cannot take in only those practices it likes from another culture; it is affected by the context of those practices, including the broader culture. Alien ideas could influence our codes of behavior and styles of social interaction, our arts and our tastes. Humans might emulate alien ways, as we rush to fads and fashions now; this impulse could be stronger if we thought we were imitating superiors. There probably would be a reaction against this, a sort of nativist movement and counterreformation combined.

**Figure 2.** A friendly alien (Klaatu) and his all-powerful robot (Gort) are attacked by paranoid soldiers in *The Day The Earth Stood Still* in 1951. *(Copyright 20th Century-Fox 1951.)*

Alien technologies and new ideas about the possible forms and purposes of economic organization could spur economic change, perhaps suggesting new opportunities for innovation and growth, or less damaging prosperity. But they also might disrupt our economies, undermining the spirit of invention and independent initiative, forcing massive readjustment and unemployment, and threatening existing economic institutions. Fear of such possibilities could provoke a new Luddite movement against alien technologies.

Encountering an alien civilization also could force us to consider more universal bases for our laws, which would encompass alien concepts as well as our own. If direct contact were to occur, we would need to adjust our conception of the legal status of non-human life forms.

## DANGERS

One of the things we tend to forget in our thinking about contact is how the aliens might react to *us.* Many scholars who have written on SETI have argued that there would be no danger in the remote contact scenario in revealing ourselves to aliens because: (1) More advanced beings would be peaceful and benign; (2) interstellar travel is so difficult and expensive that we would be insulated by distance, making direct contact impossible. These assumptions need a closer look.

Extrasolar aliens may not share the ethical standards of fairness and regard for all species. They may show no more concern for alien intelligences than we show for whales and dolphins. They may think us unintelligent. They may have had violent histories, ascending the slippery slope from barbarism to civilization several times. Their experience with competition and conflict may have instilled in them a deep concern for security. They may have had bad experiences with earlier contacts, and might—at least at first—regard us as a potential threat. Contact might come as an unpleasant surprise to a species that had believed itself to be unique and superior; learning of another technologically advanced civilization might violate the integrity of their belief system and provoke a strong reaction. Even after the communications process started, misunderstandings could provoke a nasty response. And there is the danger that Freeman Dyson is right—that we may first encounter a species in which technology is out of control, a technological cancer spreading through the Galaxy.

Contact might bring the aliens here, at least to look us over. Studies such as the British Interplanetary Society's Project Daedalus indicate that interstellar flight might be possible (though by no means easy) even for a species only slightly in advance of our own. If we are already giving serious thought to interstellar travel, it may be commonplace for more advanced beings, who might enjoy longer life spans and access to more powerful means of propulsion. Contact with us might provoke even a non-star-faring species to interstellar travel, possibly bringing eventual direct contact. Even if attack or invasion are unlikely, the aliens might wish to confine us to our own solar system, and prevent us from achieving interstellar flight, as if they were isolating a virus. That could close off human expansion, and set a final limit to our growth.

Contact with extrasolar aliens, especially a star-faring species, could be the greatest possible stimulus to the human expansion into space. Finding that another species could travel across interstellar distances would suggest that we could too; it would draw us outward, first into our solar system and then to nearby stars. We might be motivated to spread human colonies away from Earth to broaden our options for survival, should contact imply possible eventual conflict with another species. Ultimately, the existence of an alien civilization would imply a limit to our expansion, at least in one direction.

Contact might draw us into some form of interstellar politics. We would have to think about how we should relate to other cultural and political entities, and ask what role *Homo sapiens* could play in a galactic society. We must hope that relations among civilization in our galaxy are not based on some sort of interstellar social Darwinism. As a newcomer, with limited capablities to affect anything beyond near-Earth space, we might have little influence at first. Galactic geopolitics might be meaningful only if contact was with aliens whose technologies were not much better than ours.

Contact, then, could be the most important event in the history of human civilization. Its effect on us could be both positive and negative, a gigantic stimulus and a demoralizing revelation; it could stir both hopes and fears on an unprecedented scale. It could involve us in a dialogue of centuries, bringing an incalculable richness of knowledge, physical instrumentalities, and cultural growth, and opening the door to a galactic society—or it could wreck our cultures and endanger our survival. Since we are in the process of making contact more likely, we need to prepare.

ORGANIZING FOR CONTACT

Despite the popularization of the idea of extraterrestrial intelligence, we are not ready for contact. We have not created the philosophical context or the institutional framework for a calm and rational relationship with aliens. That relationship will require a broad view of the importance of life in the Universe, and of its forms and its purposes. It will require us to accept the worth of beings sprung from different evolutions. It will require political and cultural sensitivity, and tolerance for differences. It will require a long perspective on the history of our own species, and a sure knowledge of our purposes. Successfully dealing with contact will require a significant degree of consensus among human beings, and a means for expressing it.

In 1972 humanity made its first deliberate attempt to communicate with extrasolar aliens when NASA attached plaques to the Pioneer 10 and 11 spacecraft that were launched that year to swing by Jupiter before heading out of the solar system. The plaques were intended to tell any alien civilization that found them about the nature of our species and our location in the Galaxy.

Given the unlikelihood that these probes will be found in the vastness of interstellar space, the act of sending this message is more symbolic than practical. However, thinking that any contact with an extrasolar species would only be the beginning of a much larger process, I published an article in 1972 that speculated about how we might manage our relationship with an alien civilization. I argued that we could learn much by studying relations among different civilizations on Earth, and by considering the lessons of diplomatic history. I concluded that we must be as ready as possible before interstellar negotiations begin. When a group of scientists led by Frank Drake sent a powerful radio message from the Arecibo observatory in Puerto Rico in 1974, I was one of those who raised the question of what right such a small group had to speak for the entire human species without broader consultations or prior agreement.

Further developing ideas about interstellar politics in published articles over the next decade, I discovered that lawyers Andrew Haley and Ernst Fasan, among others, also had given thought to these issues. But there was no detectable interest in this subject in the world's foreign ministries, or in the United Nations.

In the absence of convincing evidence of extraterrestrial civilizations, we are unlikely to engage the sustained attention of most humans in such sweeping issues, so removed from ordinary life, or to create a permanent

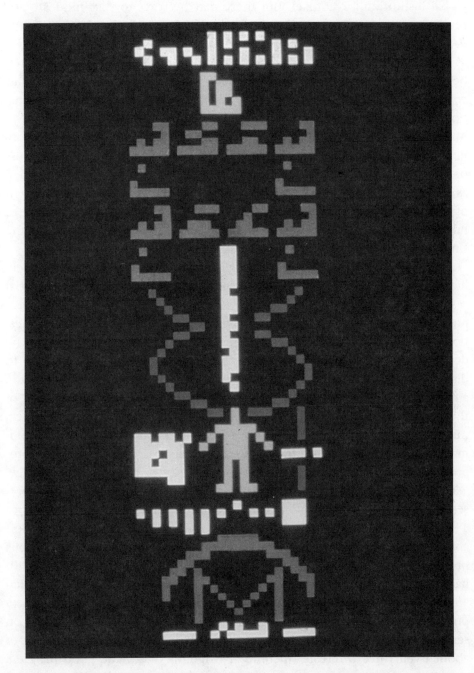

**Figure 3.** A coded message sent from the huge Arecibo radio telescope in Puerto Rico in 1974, in an experiment headed by Frank Drake. It had information about who we are and where we live.

global institution for contact. But there may be ways to start modestly, by seeking agreement among the searchers on how we would handle the *detection* of an alien civilization.

In March 1985, Professor Allen Goodman of Georgetown University began circulating drafts of a paper titled "Diplomatic Implications of Discovering Extraterrestrial Intelligence," which included a proposed international "Code of Conduct" for SETI. That code contained four principles: (1) Anyone who discovers evidence of extraterrestrial intelligence will publicly report the contact; (2) any response will be formulated by a process of international consultation; (3) visiting extraterrestrials will be regarded as envoys entitled to diplomatic immunity, protection, and aid in the event of accident; (4) in the event that extraterrestrials appear to pose a threat to human health or peace, no nation shall act without first consulting the United Nations Security Council.

At the Congress of the International Astronautical Federation in Stockholm in October 1985, John Billingham, then chief of the extraterrestrial research division at the NASA Ames Research Center in California, proposed that a session at the next Astronautical Congress address the question of international agreements on four points: The need to distribute the details of the discovery of all nations; the establishment of a mechanism to distribute this knowledge; how to determine if a response should be made and who should make the response; and how to determine the content of the response. At the October 1986 Astronautical Congress in Innsbruck, Austria, Goodman presented a revised version of his paper, titled "Diplomacy and the Search for Extraterrestrial Intelligence." It included his proposed code of conduct for relations with extraterrestrial civilizations.

Goodman and several other authors addressed the issue in papers presented at the next Astronautical Congress, in Brighton, England, in October 1987. As co-chairman of the SETI session, I noticed that there was considerable overlap among the papers. I synthesized elements from the various proposals, boiling them down to one text. As the issues associated with handling a detection appeared to be quite different from the issues associated with sending a communication, I then produced separate drafts, one a proposed agreement on detection, and the other a proposed agreement on sending a response. I presented these drafts for discussion at a session of about twenty-five interested people at Brighton. We made good progress on the detection agreement, but discussions on the communication agreement quickly bogged down in broad moral and philosophical issues.

It was clear that it would be much more feasible to reach agreement on how to handle detection than on how to handle a reply. Volunteering to act as coordinator, I circulated drafts of an agreement on the detection of extraterrestrial intelligence to interested persons over the next year, making numerous minor revisions in the text as a result of their comments but preserving its basic principles, on which correspondents generally agreed. That draft agreement was endorsed by the International Academy of Astronautics in April 1989, and by the International Institute of Space Law shortly thereafter. It also is being submitted to the International Astronautical Federation, the International Astronomical Union, and the Committee on Space Research of the International Council of Scientific Unions for their endorsements. The agreement then will be opened for signature by all of those engaged in the scientific search for extraterrestrial intelligence, hopefully in time for the planned start of NASA's expanded radio search in 1992.

The detection agreement is to be among the *searchers,* not among governments, as some institutions involved in the search, such as the Planetary Society, are not government-sponsored. Thus the agreement has no diplomatic status and is not an international agreement like the Outer Space Treaty. In fact, at the request of Czech legal scholar Vladimir Kopal, former head of the Outer Space Affairs Division of the United Nations, the agreement is now called a Declaration of Principles. The Declaration implicitly accepts an astronomical detection as the most likely scenario, but its principles could be applied to contact with another intelligent species on Earth, such as (possibly) intelligent dolphins.

The basic principles of the Declaration are those laid out by astronomer Peter Boyce in his 1987 Brighton paper: Verify the evidence in cooperation with other observers, and then tell the world. The Declaration spells out procedures for handling the detection, including the recording of the evidence and the protection of the appropriate electromagnetic wavelengths. Many of the procedures were developed by astronomer Jill Tarter, now chief scientist of NASA's SETI program. The Declaration also provides that no response to a signal or other evidence of extraterrestrial intelligence will be sent until appropriate international consultations have taken place, but leaves the mechanism of those consultations to another agreement, which could be developed from the second Brighton draft.

That draft addresses the profound questions of who should speak for Earth, and what should be said on behalf of our species. It states that communications with extraterrestrial intelligence will be undertaken on behalf of all mankind and provides that an international group will be formed to deal with the question of whether such a communication should be sent and, if it is, what its content should be. This proposed agreement is not essentially a matter of scientific research; it involves social and political questions of considerable magnitude. Refining it and gaining its acceptance by governments will be difficult. But that effort will force us to think big about our nature as a species, our shared interests, and our vision of the future.

## Declaration of Principles
## Concerning Activities Following the *Detection* of Extraterrestrial Intelligence

We, the institutions and individuals participating in the search for extraterrestrial intelligence,

Recognizing that the search for extraterrestrial intelligence is an integral part of space exploration and is being undertaken for peaceful purposes and for the common interest of all mankind,

Inspired by the profound significance for mankind of detecting evidence of extraterrestrial intelligence, even though the probability of detection may be low,

Recalling the Treaty on Principles Governing the Activities of States in the Exploration and Use of Outer Space, Including the Moon and Other Celestial Bodies, which commits states as parties to that treaty "to inform the Secretary General of the United Nations as well as the public and the international scientific community, to the greatest extent feasible and practicable, of the nature, conduct, locations and results" of their space exploration activities (Article XI),

Recognizing that any initial detection may be incomplete or ambiguous and thus require careful examination as well as confirmation, and that it is essential to maintain the highest standards of scientific responsibility and credibility,

Agree to observe the following principles for disseminating information about the detection of extraterrestrial intelligence:

1. Any individual, public or private research institution, or governmental agency that believes it has detected a signal from or other evidence of extraterrestrial intelligence (the discoverer) should seek to verify that the most plausible explanation for the evidence is the existence of extraterrestrial intelligence rather than some other natural phenomenon or anthropogenic phenomenon before making any public announcement. If the evidence cannot be confirmed as indicating the existence of extraterrestrial intelligence, the discoverer may disseminate the information as appropriate to the discovery of any unknown phenomenon.

2. Prior to making a public announcement that evidence of extraterrestrial intelligence has been detected, the discoverer should promptly inform all other observers or research organizations that are parties to this declaration, so that

those other parties may seek to confirm the discovery by independent observations at other sites and so that a network can be established to enable continuous monitoring of the signal or phenomenon. Parties to this declaration should not make any public announcement of this information until it is determined whether this information is or is not credible evidence of the existence of extraterrestrial intelligence. The discoverer should inform his/her or its relevant national authorities.

3. After concluding that the discovery appears to be credible evidence of extraterrestrial intelligence, and after informing other parties to this declaration, the discoverer should inform observers throughout the world through the Central Bureau for Astronomical Telegrams of the International Astronomical Union, and should inform the Secretary General of the United Nations in accordance with Article XI of the Treaty on Principles Governing the Activities of States in the Exploration and Use of Outer Space, including the Moon and Other Bodies. Because of their demonstrated interest in and expertise concerning the question of the existence of extraterrestrial intelligence, the discoverer should simultaneously inform the following international institutions of the discovery and should provide them with all pertinent data and recorded information concerning the evidence: the International Telecommunication Union, the Committee on Space Research, of the International Council of Scientific Unions, the International Astronautical Federation, the International Academy of Astronautics, the International Institute of Space Law and Commission 51 of the International Astronomical Union.

4. A confirmed detection of extraterrestrial intelligence should be disseminated promptly, openly, and widely through scientific channels and public media, observing the procedures in this declaration. The discoverer should have the privilege of making the first public announcement.

5. All data necessary for confirmation of detection should be made available to the international scientific community through publications, meetings, conferences, and other appropriate means.

6. The discovery should be confirmed and monitored and any data bearing on the evidence of extraterrestrial intelligence should be recorded and stored permanently to the greatest extent feasible and practicable, in a form that will make it available for further analysis and interpretation. These recordings should be made available to the international institutions listed above and to members of the scientific community for futher objective analysis and interpretation.

7. If the evidence of detection is in the form of electromagnetic signals, the parties to this declaration should seek international agreement to protect the appropriate frequencies by exercising the extraordinary procedures established within the World Administrative Radio Council of the International Telecommunication Union.

8. No response to a signal or other evidence of extraterrestrial intelligence should be sent until appropriate international consultations have taken place. The procedures for such consultations will be the subject of a separate agreement, declaration or arrangement.

9. The SETI Committee of the International Academy of Astronautics, in coordination with Commission 51 of the International Astronomical Union, will conduct a continuing review of procedures for the detection of extraterrestrial intelligence and the subsequent handling of the data. Should credible evidence of extraterrestrial intelligence by discovered, an international committee of

scientists and other experts should be established to serve as a focal point for continuing analysis of all observational evidence collected in the aftermath of the discovery, and also to provide advice on the release of information to the public. This committee should be constituted from representatives of each of the international institutions listed above and such other members as the committee may deem necessary. To facilitate the convocation of such a committee at some unknown time in the future, the SETI Committee of the International Academy of Astronautics should initiate and maintain a current list of willing representatives from each of the international institutions listed above, as well as other individuals with relevant skills, and should make that list continuously available through the Secretariat of the International Academy of Astronautics. The International Academy of Astronautics will act as the Depositary for this declaration and will annually provide a current list of parties to all the parties to this declaration.

Annex: Addresses of Institutions named in this declaration.

## ANNEX
### List of Institutions

Central Bureau for Astronomical Telegrams of the International Astronomical Union, Center for Astrophysics, 60 Garden Street, Cambridge, Massachusetts 02138, U.S.A.

Secretary-General of the United Nations, United National Headquarters, New York, New York 10017, U.S.A.

Director General, International Telecommunication Union, Place des Nations, CH-1211, Geneva-20, Switzerland

Secretary, Committee on Space Research, 51, Boulevard de Montmorency, 75015 Paris, France

Secretariat, International Astronautical Federation, 3-5 Rue Mario Nikis, 75015, Paris, France

Secretariat, International Academy of Astronautics, 3-5 Rue Mario Nikis, 75015, Paris, France

Secretariat, International Institute of Space Law, 3-5 Rue Mario Nikis, 75015, Paris, France

Secretariat, International Astronomical Union (IAU-UAI), 98 bis, Boulevard Arago, 75014 Paris, France

## Proposed Protocol
## for the Sending of Communications
## to Extraterrestrial Intelligence

The signatories agree that communications with extraterrestrial intelligence will be guided by the following principles:

1. Communications with extraterrestrial intelligence will be undertaken on behalf of all mankind, rather than specific nations, groups, or individuals.

2. Nations, organizations, and individuals will not unilaterally send communications to extraterrestrial intelligence until appropriate international consultations have taken place.

3. The signatories will not cooperate with attempts to communicate with extraterrestrial intelligence which do not conform to the principles in this protocol.

4. An international group including representation from all interested nations will be formed to deal with the question of whether such a communication should be sent and, if so, what its content should be.

5. If a decision is made to develop a communication to extraterrestrial intelligence on behalf of all mankind, the following principles will be observed:

    a. Respect for the value of life and intelligence.

    b. Respect for the value of diversity, including respect for different customs, habits, languages, creeds and religions, approaches to social organization, and styles of life.

    c. Respect for the territory and property of others.

    d. Recognition of the will to live.

    e. Recognition of the need for living space.

    f. Fair play, justice, mercy.

    g. Reciprocity and quid pro quo.

    h. Nonviolation of others.

    i. Truthfulness and non-deception.

    j. Peaceful and friendly welcome.

    k. Cooperation.

    l. Respect for knowledge, curiosity, and learning.

6. The drafters of a communication to extraterrestrial intelligence will consider detailed information about mankind to be a commodity of high value which will not be transmitted without due attention to human security and well-being, and to reciprocity.

7. In the event that extraterrestrials appear to pose a threat to human health, well-being, or peace, no nation shall act without consulting the Security Council of the United Nations.

# "ANSWER, PLEASE ANSWER"

## BY BEN BOVA

**W**e had been at the South Pole a week. The outside thermometer read fifty degrees below zero, Fahrenheit. The winter was just beginning.

"What do you think we should transmit to McMurdo?" I asked Rizzo.

He put down his magazine and half-sat up in his bunk. For a moment there was silence, except for the nearly inaudible hum of the machinery that jammed our tiny dome, and the muffled shrieking of the ever-present wind, above us.

Rizzo looked at the semi-circle of control consoles, computers, and meteorological sensors with an expression of disgust that could be produced only by a drafted soldier.

"Tell 'em it's cold, it's gonna get colder, and we've both got appendicitis and need replacements immediately."

"Very clever," I said, and started touching the buttons that would automatically transmit the sensors' memory tapes.

Rizzo sagged back into his bunk. "Why?" He asked the curved ceiling of our cramped quarters. "Why me? Why here? What did I ever do to deserve spending the whole goddammed winter at the goddammed South Pole?"

"It's strictly impersonal," I assured him. "Some bright young meteorologist back in Washington has convinced the Pentagon that the South Pole is the key to the world's weather patterns. So here we are."

"It doesn't make sense," Rizzo continued, unhearing. His dark, broad-boned face was a picture of wronged humanity. "Everybody knows that when the missiles start flying, they'll be coming over the North Pole. . . . The goddammed Army is a hundred and eighty degrees off base."

"That's about normal for the Army, isn't it?" I was a drafted soldier, too.

Rizzo swung out of the bunk and paced across the dimly-lit room. It only took a half-dozen paces; the dome was small and most of it was devoted to machinery.

"Don't start acting like a caged lion," I warned. "It's going to be a long winter."

"Yeah, guess so." He sat down next to me at the radio console and pulled a pack of cigarets from his shirt pocket. He offered one to me, and we both smoked in silence for a minute or two.

"Got anything to read?"

I grinned. "Some microspool catalogues of stars."

"Stars?"

"I'm an astronomer ... at least, I was an astronomer, before the National Emergency was proclaimed."

Rizzo looked puzzled. "But I never heard of you."

"Why should you?"

"I'm an astronomer too."

"I thought you were an electronicist."

He pumped his head up and down. "Yeah ... at the radio astronomy observatory at Greenbelt. Project OZMA. Where do you work?"

"Lick Observatory ... with the 120-inch reflector."

"Oh ... an *optical* astronomer."

"Certainly."

"You're the first optical man I've met." He looked at me a trifle queerly.

I shrugged. "Well, we've been around a few millenia longer than you static-scanners."

"Yeah, guess so."

"I didn't realize that Project OZMA was still going on. Have you had any results yet?"

It was Rizzo's turn to shrug. "Nothing yet. The project has been shelved for the duration of the emergency, of course. If there's no war, and the dish doesn't get bombed out, we'll try again."

"Still listening to the same two stars?"

"Yeah ... Tau Ceti and Epsilon Eridani. They're the only two Sun-type stars within reasonable range that might have planets like Earth."

"And you expect to pick up radio signals from an intelligent race."

"Hope to."

I flicked the ash off my cigaret. "You know, it always struck me as rather hopeless ... trying to find radio signals from intelligent creatures."

"Whattaya mean, hopeless?"

"Why should an intelligent race send radio signals out into interstellar space?" I asked. "Think of the power it requires, and the likelihood that it's all wasted effort, because there's no one within range to talk to."

"Well ... it's worth a try, isn't it ... if you think there could be intelligent creatures somewhere else ... on a planet of another star."

"Hmph. We're trying to find another intelligent race; are we transmitting radio signals?"

"No," he admitted. "Congress wouldn't vote the money for a transmitter that big."

"Exactly," I said. "We're listening, but not transmitting."

Rizzo wasn't discouraged. "Listen, the chances—just on statistical figuring alone—the chances are that there're millions of other solar systems with intelligent life. We've got to try contacting them! They might have knowledge that we don't have ... answers to questions that we can't solve yet ..."

"I completely agree," I said. "But listening for radio signals is the wrong way to do it."

"Huh?"

"Radio broadcasting requires too much power to cover interstellar distances efficiently. We should be *looking* for signals, not listening for them."

"Looking?"

"Lasers," I said, pointing to the low-key lights over the consoles. "Optical lasers. Superlamps shining out in the darkness of the void. Pump in a modest amount of electrical power, excite a few trillion atoms, and out comes a coherent, pencil-thin beam of light that can be seen for millions of miles."

"Millions of miles aren't lightyears," Rizzo muttered.

"We're rapidly approaching the point where we'll have lasers capable of lightyear ranges. I'm sure that some intelligent race somewhere in this galaxy has achieved the necessary technology to signal from star to star—by light beams."

"Then how come we haven't seen any?" Rizzo demanded.

"Perhaps we already have."

"What?"

"We're observed all sorts of variable stars—Cepheids, RR Lyrae's, T Tauri's. We assume that what we see are stars, pulsating and changing brightness for reasons that are natural, but unexplainable to us. Now, suppose what we are really viewing are laser beams, signalling from planets that circle stars too faint to be seen from Earth?"

In spite of himself, Rizzo looked intrigued.

"It would be fairly simple to examine the spectra of such light sources and determine whether they're natural stars or artificial laser beams."

"Have you tried it?"

I nodded.

"And?"

I hesitated long enough to make him hold his breath, waiting for my answer. "No soap. Every variable star I've examined is a real star."

He let out his breath in a long, disgusted puff. "Ahhh, you were kidding all along. I thought so."

"Yes," I said. "I suppose I was."

Time dragged along in the weather dome. I had managed to smuggle a small portable telescope along with me, and tried to make observations whenever possible. But the weather was usually too poor. Rizzo, almost in desperation for something to do, started to build an electronic image-amplifier for me.

Our one link with the rest of the world was our weekly radio message from McMurdo. The times for the messages were randomly scrambled, so that the chances of their being intercepted or jammed were lessened. And we were ordered to maintain strict radio silence.

As the weeks sloughed on, we learned that one of our manned satellites had been boarded by the Reds at gunpoint. Our spacecrews had put two Red automated spy-satellites out of commission. Shots had been exchanged on an ice-island in the Arctic. And six different nations were testing nuclear bombs.

We didn't get any mail of course. Our letters would be waiting for us at McMurdo when we were relieved. I thought about Gloria and our two children quite a bit, and tried not to think about the blast and fallout patterns in the San Francisco area, where they were.

"My wife hounded me until I spent pretty nearly every damned cent I had on a shelter, under the house," Rizzo told me. "Damned shelter is fancier than the house. She's the social leader of the disaster set. If we don't have a war, she's gonna feel damned silly."

I said nothing.

The weather cleared and steadied for a while (days and nights were indistinguishible during the long Antarctic winter) and I split my time evenly between monitoring the meteorological sensors and observing the stars. The snow had covered the dome completely, of course, but our "snorkel" burrowed through it and out into the air.

"This dome's just like a submarine, only we're submerged in snow instead of water," Rizzo observed. "I just hope we don't sink to the bottom."

"The calculations show that we'll be all right."

He made a sour face. "Calculations proved that airplanes would never get off the ground."

The storms closed in again, but by the time they cleared once more, Rizzo had completed the image-amplifier for me. Now, with the tiny telescope I had, I could see almost as far as a professional instrument would allow. I could even lie comfortably in my bunk, watch the amplifier's viewscreen, and control the entire set-up remotely.

Then it happened.

At first it was simply a curiosity. An oddity.

I happened to be studying a Cepheid variable star—one of the huge, very bright stars that pulsate so regularly that you can set your watch by them. It had attracted my attention because it seemed to be unusually close for a Cepheid—only 700 lightyears away. The distance could be easily gauged by timing the star's pulsations.*

I talked Rizzo into helping me set up a spectrometer. We scavenged shamelessly from the dome's spare parts bin and finally produced an instrument that would break up the light of the star into its component wavelengths, and thereby tell us much about the star's chemical composition and surface temperature.

At first I didn't believe what I saw.

The star's spectrum—a broad rainbow of colors—was criss-crossed with narrow dark lines. That was all right. They're called absorption lines; the Sun has thousands of them in its spectrum. But one line—*one*—was an

---

*Astronomers have been able, since about 1910, to estimate the distances of Cepheid variable stars by timing their pulsations. The length of this type of star's pulsation is a true measure of its intrinsic brightness. Comparing the star's actual brightness to its apparent brightness, as seen from Earth, gives a good value for the star's distance.*

insolently bright emission line. All the laws of physics and chemistry said it couldn't be there.

But it was.

We photographed the star dozens of times. We checked our instruments ceaselessly. I spent hours scanning the star's "official" spectrum in the microspool reader. The bright emission line was not on the catalogue spectrum. There was nothing wrong with our instruments.

Yet the bright line showed up. It was real.

"I don't understand it," I admitted. "I've seen stars with bright emission spectra before, but a single bright line in an absorption spectrum! It's unheard-of. One single wavelength . . . one particular type of atom at one precise energy-level . . . why? Why is it emitting energy when the other wavelengths aren't?"

Rizzo was sitting on his bunk, puffing a cigaret. He blew a cloud of smoke at the low ceiling. "Maybe it's one of those laser signals you were telling me about a couple weeks ago."

I scowled at him. "Come on, now. I'm serious. This thing has me puzzled."

"Now wait a minute . . . you're the one who said radio astronomers were straining their ears for nothing. You're the one who said we ought to be looking. So look!" He was enjoying his revenge.

I shook my head, and turned back to the meteorological equipment.

But Rizzo wouldn't let up. "Suppose there's an intelligent race living on a planet near a Cepheid variable star. They figure that any other intelligent creatures would have astronomers who'd be curious about their star, right? So they send out a laser signal that matches the star's pulsations. When you look at the star, you see their signal. What's more logical?"

"All right," I groused. "You've had your joke . . ."

"Tell you what," he insisted. "Let's put that one wavelength into an oscilloscope and see if a definite signal comes out. Maybe it'll spell out 'Take me to your leader' or something."

I ignored him and turned my attention to Army business. The meteorological equipment was functioning perfectly, but our orders read that one of us had to check it every twelve hours. So I checked and tried to keep my eyes from wandering as Rizzo tinkered with a photocell and oscilloscope.

"There we are," he said, at length. "Now let's see what they're telling us."

In spite of myself I looked up at the face of the oscilloscope. A steady, gradually sloping greenish line was traced across the screen.

"No message," I said.

Rizzo shrugged elaborately.

"If you leave the 'scope on for two days, you'll find that the line makes a full swing from peak to null," I informed him. "The star pulsates every two days, bright to dim."

"Let's turn up the gain," he said, and he flicked a few knobs on the front of the 'scope.

The line didn't change at all.

"What's the sweep speed?" I asked.

"One nanosecond per centimeter." That meant that each centimeter-wide square on the screen's face represented one billionth of a second. There are as many nonoseconds in one second as there are seconds in thirty-two years.

"Well, if you don't get a signal at that sensitivity, there just isn't any signal there," I said.

Rizzo nodded. He seemed slightly disappointed that his joke was at an end. I turned back to the meteorological instruments, but I couldn't concentrate on them. Somehow I felt disappointed, too. Subconsciously, I suppose, I had been hoping that Rizzo actually would detect a signal from the star. *Fool!* I told myself. But what could explain that bright emission line? I glanced up at the oscilloscope again.

And suddenly the smooth steady line broke into a jagged series of millions of peaks and nulls!

I stared at it.

Rizzo was back on his bunk again, reading one of his magazines. I tried to call him, but the words froze in my throat. Without taking my eyes from the flickering 'scope, I reached out and touched his arm.

He looked up.

"Holy Mother of God," Rizzo whispered.

For a long time we stared silently at the fluttering line dancing across the oscilloscope screen, bathing our tiny dome in its weird greenish light. It was eerily fascinating, hypnotic. The line never stood still; it jabbered and stuttered, a series of millions of little peaks and nulls, changing almost too fast for the eye to follow, up and down, calling to us, speaking to us, up, down, never still, never quiet, constantly flickering its unknown message to us.

"Can it be . . . people?" Rizzo wondered. His face, bathed in the greenish light, was suddenly furrowed, withered, ancient: a mixture of disbelief and fear.

"What else could it be?" I heard my own voice answer. "There's no other explanation possible."

We sat mutely for God knows how long.

Finally Rizzo asked, "What do we do now?"

The question broke our entranced mood. What do we do? What action do we take? We're thinking men, and we've been contacted by other creatures that can think, reason, send a signal across seven hundred lightyears of space. So don't just sit there in stupified awe. Use you're brain, prove that you're worthy of the tag *sapiens*.

"We decode the message," I announced. Then, as an afterthought, "But don't ask me how."

We should have called McMurdo, or Washington. Or perhaps we should have attempted to get a message through to the United Nations. But we never even thought of it. This was our problem. Perhaps it was the sheer isolation of our dome that kept us from thinking about the rest of the world. Perhaps it was sheer luck.

"If they're using lasers," Rizzo reasoned, "they must have a technology something like ours."

"Must *have had*," I corrected. "That message is seven hundred years old, remember. They were playing with lasers when King John was signing the Magna Charta and Genghis Khan owned most of Asia. Lord knows what they have now."

Rizzo blanched and reached for another cigaret.

I turned back to the oscilloscope. The signal was still flashing across its face.

"They're sending out a signal," I mused, "probably at random. Just beaming it out into space, hoping that someone, somewhere will pick it up. It must be in some form of code . . . but a code that they feel can be easily cracked by anyone with enough intelligence to realize that there's a message there."

"Sort of an interstellar Morse code."

I shook my head. "Morse code depends on both sides knowing the code. There's no key."

"Cryptographers crack codes."

"Sure. If they know what language is being used. We don't know the language, we don't know the alphabet, the thought processes . . . nothing."

"But it's a code that can be cracked easily," Rizzo muttered.

"Yes," I agreed. "Now what the hell kind of a code can they assume will be known to another race that they've never seen?"

Rizzo leaned back on his bunk and his face was lost in shadows.

"An interstellar code," I rambled on. "Some form of presenting information that would be known to almost any race intelligent enough to understand lasers ..."

"Binary!" Rizzo snapped, sitting up on the bunk.

"What?"

"Binary code. To send a signal like this, they've gotta be able to write a message in units that're only a billionth of a second long. That takes computers. Right? Well, if they have computers, they must figure that we have computers. Digital computers run on binary code. Off or on ... go or no-go. It's simple. I'll bet we can slap that signal on a tape and run it through our computer here."

"To assume that they use computers exactly like ours ..."

"Maybe the computers are completely different," Rizzo said excitedly, "but the binary code is basic to them all. I'll bet on that! And this computer we've got here—this transistorized baby—she can handle more information than the whole Army could feed into her. I'll bet nothing has been developed anywhere that's better for handling simple one-plus-one types of operations."

I shrugged. "All right. It's worth a trial."

It took Rizzo a few hours to get everything properly set up. I did some arithmetic while he worked. If the message was in binary code, that meant that every cycle of the signal—every flick of the dancing line on our screen—carried a bit of information. The signal's wavelength was 5000 Angstroms; there are a hundred million Angstrom units to the centimeter; figuring the speed of light ... the signal could carry, in theory at least, something like 600 trillion bits of information *per second.*

I told Rizzo.

"Yeah, I know. I've been going over the same numbers in my head." He set a few switches on the computer control board. "Now let's see how many of the 600 trillion we can pick up." He sat down before the board and pressed a series of buttons.

We watched, hardly breathing, as the computer's spools began spinning and the indicator lights flashed across the control board. Within a few minutes, the printer chugged to life.

Rizzo swivelled his chair over to the printer and held up the unrolling sheet in a trembling hand.

Numbers. Six digit numbers. Completely meaningless.

"Gibberish," Rizzo snapped.

It was peculiar. I felt relieved and disappointed at the same time.

"Something's screwy," Rizzo said. "Maybe I fouled up the circuits . . ."

"I don't think so," I answered. "After all, what did you expect out of the computer? Shakespearean poetry?"

"No, but I expected numbers that would make some sense. One and one, maybe. Something that means something. This stuff is nowhere."

Our nerves must have really been wound tight, because before we knew it we were in the middle of a hasty argument—and it was over nothing, really. But in the middle of it:

"Hey, look," Rizzo shouted, pointing to the oscilloscope.

The message had stopped. The 'scope showed only the calm, steady line of the star's basic two-day-long pulsation.

It suddenly occurred to us that we hadn't slept for more than 36 hours, and we were both exhausted. We forgot the senseless argument. The message was ended. Perhaps there would be another; perhaps not. We had the telescope, spectrometer, photocell, oscilloscope, and computer set to record automatically. We collapsed into our bunks. I suppose I should have had monumental dreams. I didn't. I slept like a dead man.

When we woke up, the oscilloscope trace was still quiet.

"Y'know," Rizzo muttered, "it might just be a fluke . . . I mean, maybe the signals don't mean a damned thing. The computer is probably translating nonsense into numbers just because it's built to print out numbers and nothing else."

"Not likely," I said. "There are too many coincidences to be explained. We're receiving a message, I'm certain of it. Now we've got to crack the code."

As if to reinforce my words, the oscilloscope trace suddenly erupted into the same flickering pattern. The message was being sent again.

We went through two weeks of it. The message would run through for seven hours, then stop for seven. We transcribed it on tape 48 times and ran it through the computer constantly. Always the same result—six-digit numbers; millions of them. There were six different seven-hour-long messages, being repeated one after the other, constantly.

We forgot the meteorlogical equipment. We ignored the weekly messages from McMurdo. The rest of the world became a meaningless fiction to us. There was nothing but the confounded, tantalizing, infuriating, enthralling message. The National Emergency, the bomb tests, families, duties—

all transcended, all forgotten. We ate when we thought of it and slept when we couldn't keep our eyes open any longer. The message. What was it? What was the key to unlock its meaning?

"It's got to be something universal," I told Rizzo. "Something universal . . . in the widest sense of the term."

He looked up from his desk, which was wedged in between the end of his bunk and the curving dome wall. The desk was littered with printout sheets from the computer, each one of them part of the message.

"You've only said that a half-million times in the past couple weeks. What the hell is universal? If you can figure that out, you're damned good."

*What is universal?* I wondered. *You're an astronomer. You look out at the universe. What do you see?* I thought about it. *What do I see? Stars, gas, dust clouds, planets . . . what's universal about them? What do they all have that . . .*

"Atoms!" I blurted.

Rizzo cocked a weary eye at me. "Atoms?"

"Atoms. Elements. Look . . ." I grabbed up a fistful of the sheets and thumbed through them. "Look . . . each message starts with a list of numbers. Then there's a long blank to separate the opening list from the rest of the message. See? Every time, the same length list."

"So?"

"The periodic table of the elements!" I shouted into his ear. "That's the key!"

Rizzo shook his head. "I thought of that two days ago. No soap. In the first place, the list that starts each message isn't always the same. It's the same length, all right, but the numbers change. In the second place, it always begins with 100000. I looked up the atomic weight of hydrogen— it's 1.008 something."

That stopped me for a moment. But then something clicked into place in my mind.

"Why is the hydrogen weight 1.008?" Before Rizzo could answer, I went on, "For two reasons. The system we use arbitrarily rates oxygen as 16-even. Right? All the other weights are calculated from oxygen's. And we also give the average weight of an element, counting all its isotopes. Our weight for hydrogen also includes an adjustment for tiny amounts of deuterium and tritium. Right? Well, suppose *they* have a system that rates hydrogen as a flat one: 1.00000. Doesn't *that* make sense?"

"You're getting punchy," Rizzo grumbled. "What about the isotopes?

How can they expect us to handle decimal points if they don't tell us about them ... mental telepathy? What about ..."

"Stop arguing and start calculating," I snapped. "Change that list of numbers to agree with our periodic table. Change 1.00000 to 1.008-whatever-it-is and tackle the next few elements. The decimals shouldn't be so hard to figure out."

Rizzo grumbled to himself, but started working out the calculations. I stepped over to the dome's microspool library and found an elementary physics text. Within a few minutes, Rizzo had some numbers and I had the periodic table focused on the microspool reading machine.

"Nothing," Rizzo said, leaning over my shoulder and looking at the screen. "They don't match at all."

"Try another list. They're not all the same."

He shrugged and returned to his desk. After a while he called out, "their second number is 3.97123; it works out to 4.003-something."

It checked! "Good. That's helium. What about the next one, lithium?"

"That's 6.940."

"Right!"

Rizzo went to work furiously after that. I pushed a chair to the desk and began working up from the end of the list. It all checked out, from hydrogen to a few elements beyond the artificial ones that had been created in the laboratories here on Earth.

"That's it," I said. "That's the key. That's our Rosetta Stone ... the periodic table."

Rizzo stared at the scribbled numbers and jumble of papers. "I bet I know what the other lists are ... the ones that don't make sense."

"Oh?"

"There are other ways to identify the elements ... vibration resonances, quantum wave-lengths ... somebody named Lewis came out a couple years ago with a Quantum Periodic Table ..."

"They're covering all the possibilities. There are messages for many different levels of understanding. We just decoded the simplest one."

"Yeah."

I noticed that as he spoke, Rizzo's hand—still tightly clutching the pencil—was trembling and white with tension.

"Well?"

Rizzo licked his lips. "Let's get to work."

We were like two men possessed. Eating, sleeping, even talking was ignored completely as we waded through the hundreds of sheets of paper.

We could decode only a small percentage of them, but they still represented many hours of communication. The sheets that we couldn't decode,
we suspected, were repetitions of the same message that we were working
on.

We lost all concept of time. We must have slept, more than once, but I
simply don't remember. All I can recall is thousands of numbers, row upon
row, sheet after sheet of numbers ... and my pencil scratching symbols of
the various chemical elements over them until my hand was so cramped I
could no longer open the fingers.

The message consisted of a long series of formulas; that much was
certain. But, without punctuation, with no knowledge of the symbols that
denote even such simple things as "plus" or "equals" or "yields," it took us
more weeks of hard work to unravel the sense of each equation. And even
then, there was more to the message than met the eye.

"Just what the hell are they driving at?" Rizzo wondered aloud. His face
had changed: it was thinner, hollow-eyed, weary, covered with a scraggly
beard.

"Then you think there's a meaning behind all these equations, too?"

He nodded. "It's a message, not just a contact. They're going to an awful
lot of trouble to beam out this message, and they're repeating it every seven
hours. They haven't added anything new in the weeks we've been watching."

"I wonder how many years or centuries they've been sending out this
message, waiting for someone to answer them."

"Maybe we should call Washington ..."

"No!"

Rizzo grinned. "Afraid of breaking radio silence?"

"Hell no. I just want to wait until we're relieved, so we can make this
announcement in person. I'm not going to let some old wheezer in Washington get credit for this.... Besides, I want to know just what they're
trying to tell us."

It was agonizing, painstaking work. Most of the formulas meant nothing
to either one of us. We had to ransack the dome's meager library of
microspools to piece them together. They started simply enough—basic
chemical combinations: carbon and two oxygens yield $CO_2$; two hydrogens
and oxygen give water. A primer ... not of words, but of equations.

The equations became steadily longer and more complex. Then, abruptly,
they simplified, only to begin a new deepening, simplify again, and finally
become very complicated just at the end. The last few lines were obviously
repetitious.

Gradually, their meaning became clear to us.

The first set of equations started off with simple, naturally occurring energy yielding formulas. The oxidation of cellulose (we found the formula for that in an organic chemistry text left behind by one of the dome's previous occupants), which probably referred to the burning of plants and vegetation. A string of formulas that had groupings in them that I dimly recognized as amino acids—no doubt something to do with digesting food. There were many others, including a few that Rizzo claimed had the expression for chlorophyll in them.

"Naturally occurring, energy-yielding reactions," Rizzo summarized. "They're probably trying to describe the biological set-up on their planet."

It seemed an inspired guess.

The second set of equations again began with simple formulas. The cellulose-burning reaction appeared again, but this time it was followed by equations dealing with the oxidation of hydrocarbons: coal and oil burning? A long series of equations that bore repeatedly the symbols for many different metals came up next, followed by more on hydrocarbons, and then a string of formulas that we couldn't decipher at all.

This time it was my guess: "These look like energy-yielding reactions, too. At least in the beginning. But they don't seem to be naturally occurring types. Then comes a long story about metals. They're trying to tell us the history of their technological development—burning wood, coal and eventually oil; smelting metals . . . they're showing us how they developed their technology."

The final set of equations began with an ominous simplicity: a short series of very brief symbols that had the net result of four hydrogen atoms building into a helium atom. Nuclear fusion.

"That's the proton-proton reaction," I explained to Rizzo. "The type of fusion that goes on in the Sun."

The next series of equations spelled out the more complex carbon-nitrogen cycle of nuclear fusion, which was probably the primary energy source of their own Cepheid variable star. Then came a long series of equations that we couldn't decode in detail, but the symbols for uranium and plutonium, and some of the heavier elements, kept cropping up.

Then came one line that told us the whole story: the lithium-hydride equation—nuclear fusion bombs.

The equations went on to more complex reactions, formulas that no man on Earth had ever seen before. They were showing us the summation of their knowledge, and they had obviously been dealing with nuclear energies for much longer than we have on Earth.

But interspersed among the new equations, they repeated a set of formulas that always began with the lithium hydride fusion reaction. The message ended in a way that wrenched my stomach: the fusion bomb reaction and its cohorts were repeated ten straight times.

I'm not sure of what day it was on the calendar, but the clock on the master control console said it well past eleven.

Rizzo rubbed a weary hand across his eyes. "Well, what do you think?"

"It's pretty obvious," I said. "They have the bombs. They've had them for quite some time. They must have a lot of other weapons, too—more . . . advanced. They're trying to tell us their history with the equations. First they depended on natural sources of energy, plants and animals; then they developed artificial energy sources and built up a technology; finally they discovered nuclear energy."

"How long do you think they've had the bombs?"

"Hard to tell. A generation . . . a century. What difference does it make? They have them. They probably thought, at first, that they could learn to live with them . . . but imagine what it must be like to have those weapons at your fingertips . . . for a century. Forever. Now they're so scared of them that they're beaming their whole history out into space, looking for someone to tell them how to live with the bombs, how to avoid using them."

"You could be wrong," Rizzo said. "They could be boasting about their arsenal."

"Why? For what reason? No . . . the way they keep repeating those last equations. They're pleading for help."

Rizzo turned to the oscilloscope. It was flickering again.

"Think it's the same thing?"

"No doubt. You're taping it anyway, aren't you?"

"Yeah, sure. Automatically."

Suddenly, in mid-flight, the signal winked off. The pulsations didn't simply smooth out into a steady line, as they had before. The screen simply went dead.

"That's funny," Rizzo said, puzzled. He checked the oscilloscope. "Nothing wrong here. Something must've happened to the telescope."

Suddenly I knew what had happened. "Take the spectrometer off and turn on the image-amplifier," I told him.

I knew what we would see. I knew why the oscilloscope beam had suddenly gone off scale. And the knowledge was making me sick.

Rizzo removed the spectrometer set-up and flicked the switch that energized the image-amplifier's viewscreen.

"Holy God!"

The dome was flooded with light. The star had exploded.

"They had the bombs all right," I heard myself saying. "And they couldn't prevent themselves from using them. And they had a lot more, too. Enough to push their star past its natural limits."

Rizzo's face was etched in the harsh light.

"I've gotta get out of here," he muttered, looking all around the cramped dome. "I've gotta get back to my wife and find someplace where it's safe . . ."

"Someplace?" I asked, staring at the screen. "Where?"

# CHAPTER 9

# HOW TO PARTICIPATE IN SETI

*We are accustomed to thinking of research efforts such as SETI in terms of large teams of professional scientists using the latest state-of-the-art hardware and software. Yet it is possible for individuals—dedicated, inquisitive amateurs—to help search the skies for the first signals from an extraterrestrial intelligence.*

*You can be part of SETI. Using modest, easily available equipment, you can join the search for extraterrestrial intelligence. In this chapter, NASA scientist Kent Cullers and astronomer William Alschuler explain "backyard SETI."*

*Amateur astronomers have made important contributions to the discovery of comets and asteroids. Ham-radio operators have helped to track artificial satellites and deep space probes. Pioneering research in the fields of electricity, meteorology, and heat flow was done by an amateur scientist named Benjamin Franklin. The race to find the first ETI signal is not entirely the province of professional astronomers with big telescopes. You can help, using equipment that is commercially available. Even if you don't make the first contact, you will have participated in the noble endeavor.*

# INDIVIDUAL INVOLVEMENT

## BY D. KENT CULLERS
## AND WILLIAM R. ALSCHULER

It is 5:00 A.M. Atlantic Standard Time. From the radar console where I sit writing this, the raucous jungle noises are nearly inaudible. Instead of these, the air conditioning hums, cooling klystrons and computers below the temperature of the tropical night.

If I were to take a ten-minute stroll, I could stand on a support platform five hundred feet above twenty acres of metal mesh, the largest radio antenna in the world. At this moment, whatever signals are falling on this giant antenna are unheeded. The control room is silent. Only the control operator and I are present. Normally, this room, with racks of receivers, amplifiers, and processing equipment is a beehive of activity. The Arecibo antenna performs radio astronomy experiments year round. People from all over the world requiring the extremely high sensitivity of this instrument propose research, and the telescope time is in great demand.

Right now, however, something unusual, though not unprecedented, has occurred. An experiment has finished early. Part of the team scheduled to observe next, I am unable to sleep. The other team members are more sensible after a twenty-four-hour period of uninterrupted observing and data processing. They are sleeping so that when our new run starts in an hour, at least someone will be thinking straight. But now, I have just been given sole use of the telescope for the next hour. It is an awesome thought that I could do anything my heart desires. Unfortunately, I have no idea where to point the telescope to find an extraterrestrial signal. So, I begin checking the set-up we plan to use in an hour. Double- and triple-checking prevents mistakes.

You have already read the arguments for the plausibility of extraterrestrial intelligence. These arguments are not proof. The scientific revolution has swept away countless attempts to deduce the nature of physical reality from first causes. Physics is an experimental science. Either communicating ETs exist or they don't. The debate on this subject will be settled by observation, not by logical deduction.

Of course, those, like me, who are devoting their lives to a search for ET, believe that there is a good possibility of success given enough time and resources. Basically, this belief is why I am presently observing at Arecibo. The team gathered here is taking the first steps in learning how to carry out an automated search of the sky. I am the NASA SETI signal-detection team leader. My job is to tell the difference between a signal from ET and anything else that might look similar.

Two phenomena concern me on this particular night. First, radio tele-scope optical telescopes, are susceptible to local interference. In the case of optical instruments, this may be light from nearby cities, which scatters into the field of view when the telescope is pointed at a star. Radio telescopes have much the same problem. Local radio transmitters, from taxicab radios to communications satellites, send stray signals into the Arecibo dish. Not only is the interference likely to be strong, but the signals have the systematic characteristics and patterns expected of intelligence. Consequently, they will pass the tests that should find a transmission from an extraterrestrial technology and discriminate against natural cosmic noise. How severe is this interference at our most important telescope site? At the moment, I am collecting the data to find out. The analysis will take months.

Second, even if all interference from the local environment is success-fully excised, noise can fool you once in a while. Noise is a random process. In searching the sky for five years over a wide band of frequencies, the total amount of data processed will be thousands of terabits (a terabit is $10^{12}$ bits). Each second, NASA instruments will process the information equiva-lent of several entire *Encyclopedia Britannicas* (several gigabits—$10^9$ bits—per second). This information, almost entirely cosmic noise, will be, by analogy, random letter combinations. Sometimes, however, just by chance, a noise "word" will look like an intelligent signal. In principle, complete elimination of false alarms from random noise is impossible. However, I must know the expected false alarm rate for my system. So far, on this observing run, the false alarm rate is that predicted from the theory.

So what I'm doing is learning how to conduct a massive search of the sky, a look at many frequencies and many stars. It will begin in the middle 1990s and continue for about seven years. It will use the world's largest antennas. It will be a billion times larger than any search before. It will cost $100,000,000. In the face of this, what can an individual do?

Actually, individuals from the interested public can help a lot. Politically, a few personal letters supporting the idea of SETI have disproportionate leverage with Congress. This should not surprise you. The much more surprising fact is that skilled amateurs, using home-built equipment, can actually make technical contributions to the SETI effort. They can search at frequencies and for signal types that NASA will not. They can more easily test new ideas than NASA can in a large program. They can attain detection sensitivies comparable to those NASA will have in the same search domain. It is even possible that an amateur proposal of exceptional merit can get time at a major observatory using state-of-the-art equipment. The rest of this chapter will tell you why this is possible and how you can start your own SETI program.

## HOW "BIG" IS A SEARCH

If you are not an expert on radio, more particularly radio astronomy, but would like to plunge in, this section is for you. I will try to keep the math at a minimum, but many of the concepts are technical and require some thought. If you trust me utterly, dangerous at the best of times, you can skip to the end and accept the conclusions. For those of you still with me, here goes. I promise you that by the end we will have proved together that an individual with about three thousand dollars can search a small region of frequencies and a few stars more sensitively than has any ETI search conducted to date. In fact, it is within your means to look in a direction of your choosing and at a small band of frequencies more sensitively than will any search currently planned. This carries with it, however, a vitally important caveat. You must know where to look. To rephrase Lincoln, "You can beat the professionals in some of the sky some of the time, but it's impossible in all of the sky all of the time."

People like me in the SETI biz talk glibly of "search space," "nine-dimensional hay stacks," and "terabits" to describe the size of their endeavors. Usually, after only an hour or two of argument, we can agree on what we mean. The fact that two people never use exactly the same numbers to describe the size of a particular ETI search shows how many hidden assumptions go into evaluating the effectiveness of methods for finding ETI. Let me explain this by analogy with visually observing the sky. On a clear, dark night, how many stars can you see? In the visible Universe there are at least a million trillion stars. The unaided eye can see about two thousand of these in good conditions from any one place on Earth, at any one time. Primarily, this small number is due to the fact that most stars fall below the eye's detection threshold. A similar effect exists in radio astronomy, and its cause is easy to understand.

For a signal to be detected, it must stand out against the background of receiver noise that is constantly present in any radio equipment. All objects in the Universe have a finite temperature, which is a measure of the energy in random particle motions. The important particles in your radio set are the electrons. Their average motion is the current which, when amplified, is the usable signal output. Finite temperature means that random electron motions will cause fluctuations in the receiver's output. If you connect a speaker to this output you hear noise called static. If you point your antenna to the sky you will, if it is sensitive enough, also hear cosmic static and other natural radio signals.

If an incoming signal is much weaker than this thermal noise, it is not noticed against the background roar. If it is much stronger, it is obvious. Thus, we define the detection threshold for a signal, rather intuitively, as that level where the signal power is equal to the noise power. A signal becomes more detectable if we increase the signal power or if we cut the noise power. The important thing is their ratio, often called the SNR, signal-to-noise ratio.

Obviously, we want to search as many stars as possible when looking for ETs. We cannot control the power of their transmitters. We can, however, decrease the noise in our receivers. One way is to cool the receiving equipment, thus decreasing the random electron motions. Another way is to divide the incoming data into small pieces, either in frequency or in time. The noise in each small piece of data is tiny compared to the noise in all the data taken together. If there is a signal in one piece, concentrated there, it will be more detectable in that segment alone than it would be against the noise in the entire data stream. The largest planned SETI endeavor, NASA's effort, uses knowledge of signal processing to carve up the frequency and time dimensions in pieces likely to contain large signals but little noise. This search concentrates on continuosly present or regularly recurring simple signals. Such searches are good at finding certain types of regularly pulsed radar signals as well as amplitude-modulated radio (AM) and TV transmissions with a narrow carrier wave component. Other transmission types, such as commercial FM stereo, are very poorly matched to NASA's search. Since the information in an FM signal is represented as frequency changes, the signal covers many channels in an irregular pattern determined by the transmission's content. If ETs only broadcast in FM stereo, NASA is unlikely to find the signal.

The point is that even the largest planned computer processing systems on the largest available antennas, looking at the quietest frequencies for interstellar transmissions, cannot do it all. A lucky amateur could scoop the pros. In a later section, I will tell you about what the big searches are doing so you can see what they are missing.

The one obvious advantage that professional astronomers have is antenna size or collecting area. Basically, this is why professionals can mount bigger searches. The other advantage is information processing power. Bigger computers can look at more frequencies, wider bandwidths, for more signal types, in the same length of time (but amateurs may be able to slip in with unusual processing techniques; computer power is relatively cheap). Most scientists feel that the probability of detecting ETI is proportional to the number of stars searched and to the frequency range covered. This is only roughly true, however. For example, most astronomers feel that we should concentrate on stars like the Sun, stars that live long enough for life as we know it to evolve and that are relatively warm, maintaining a life-sustaining liquid element on nearby planets. Further, some frequencies, where natural noise is high or where local Earth-based signals obscure weak signals, are not likely to turn up ETI and professionals will ignore them. Nevertheless, assume that the probability of finding ETI is proportional to the number of stars and frequencies covered.

The distance to which a signal is detectable, keeping everything like the receiver noise and incoming signal power constant, is proportional to the diameter of an antenna. The signal received by an antenna of constant size decreases with source distance, just like the force of gravity. If the distance is doubled between transmitter and receiver, the signal is decreased by a factor of four. This is the inverse square law at work. It applies because by the time the signal reaches the receiver from twice the distance, the same amount of transmitted power has spread over four times the area. The receiving antenna, still the same size, collects one quarter of the signal it had before.

If a source's distance is doubled and the diameter of the antenna is doubled, the received signal strength remains constant. The signal per unit area is one quarter as great after the doubling, but there is four times as much antenna area to soak up the transmission. Thus, by doubling the distance and doubling the antenna size, the signal is just as detectable as before. Therefore, the detection range is proportional to the antenna diameter.

It is a good approximation to assume that the number of stars within a given distance is proportional to the cube of that distance. In other words, in our region of space, the number of stars is proportional to the volume we can see. Obviously, on large scales this is not true. The Galaxy is shaped like a convex lens, not a sphere, so that in some directions volumes can be found that are almost empty of stars. For distances up to a thousand light-years, however, the assumption is good. Since the distance we can see out to is proportional to the antenna diameter, the number of stars is proportional to the antenna diameter, cubed.

If we accept the assumption that the probability of finding ETI is proportional to the stars and frequencies searched, we get:

$$p = kfd^3$$

where $p$ is the probability of finding ETI, $k$ is a constant known by no man (it is related to the result of the Drake Equation), $f$ is the frequency range searched, and $d$ is the antenna diameter. We are in the odd situation of knowing what is better—big antennas, lots of frequency coverage—without actually knowing what the probability of finding ETI really is. Nonetheless, we now know enough to see if an amateur can do anything useful in this age of professional ETI hunters.

The antenna at Arecibo, Puerto Rico, is three hundred meters across. A typical amateur setup might use a satellite dish three meters across, one hundred times smaller. An individual can afford a PC to process his data. A professional can get a Cray-sized supercomputer, which runs about one thousand times faster. Frequency coverage is essentially proportional to the speed of data processing.

So, if you substitute into the above equation you will see that the professional has a billion times ($100^3 \times 1000$) the probability of the amateur of finding ETI. This is true if you accept all the prior underlying assumptions. (In the data-rich world of the near future, it will be theoretically possible to test your favorite algorithm on data direct from professional radio telescopes, sent to you via satellite or, better yet, fiber optic link. In this way, you might overcome the antenna disadvantage of amateur searches. At present, NASA's SETI project has no money budgeted to make raw data available outside the observatory.)

Is it possible, though, that a billion-to-one long shot could pay off in favor of an amateur being the first discoverer of ETI? I believe it is just possible, and I will explain why. Though professional searches cover much wider frequency ranges than amateur efforts, they do not cover all good frequencies. First of all, the really big NASA search hasn't started yet. It will not be of significant size until the middle '90s. It will cover frequencies for nearby Sun-like stars between one and three billion cycles per second (1–3 GHz) using Arecibo and other large antennas, and for the whole sky with smaller antennas over frequencies between one and ten billion cycles per second. It will be sensitive to continuous signals like those of TV and to pulses like those of some radars. It will not see FM stereo, many types of military transmissions, or TV satellites, which also use frequency modulation.

Many of our strongest Earth-based transmissions currently take place below a billion cycles per second (1 GHz). FM transmissions occur at about one hundred million cycles per second. TV carriers occupy channels up to the highest UHF allocation at eight hundred million cycles per second. The professionals have ruled these out, at least for the time being, because most radio telescopes are not efficient at these frequencies and because the cosmic noise is slightly higher there than at the noise minimum. However, if you are an amateur in a radio-quiet location or you can arrange efficient interference filtering, you might want to search them. Good automated scanning receivers that cover them cost less than one thousand dollars. Remember that you are likely, unless you take precautions, to detect Earth-based signals instead of ETs.

So perhaps, you admit, there are some good frequencies left for prospecting. What about the obvious professional antenna advantage?

All of the stars cannot be observed with the antenna at Arecibo, Puerto Rico. That antenna uses an unusual construction that makes it cheap to build, but which allows effective use only near the vertical. Thus, though a band of stars is visible as the Earth turns, only those near the celestial equator (which is almost overhead in Puerto Rico) can be seen. In all, only about one-third of the Sun-like stars can be viewed by Arecibo. The rest of the stars will be observed at other latitudes by telescopes ranging in size from thirty to sixty meters in diameter. We have fairly complete lists of Sun-like stars only to about one hundred light-years distance. Beyond this distance, stars like the Sun are too dim for us to be certain that they have been included in our data bases. (The Sun would be about 7.5 magnitude, just too faint for the naked eye, at one hundred light-years.) The current plan is to use the biggest available radio telescopes to look for a long time at the approximately one thousand nearby Sun-like stars that we know about. The rest of the sky, most of it, will be swept quickly by antennas having a diameter of thirty-four meters. This allows a complete, if less sensitive, look at the sky.

The professional-antenna advantage over our amateur set-up using a three-meter dish is typically not one to a million, but something like one to ten thousand. Things are looking better. There is, however, another surprise. If he knows the frequency of ETI, a *big* if, an amateur can purchase software that does better signal processing than that planned by the professionals at NASA-Ames and JPL. This can increase amateur sensitivities for some particular star and frequency so that his antenna disadvantage is overcome completely. In some small part of the planned search, an amateur can do a professional job. Many amateurs, working cooperatively, may do even better.

## SENSITIVITY OF THE SEARCH

This is, by far, the most technical section in this chapter. It is my area of expertise, and by the end, some of you will undoubtedly say that I am lost in the details of my job. Unfortunately, the details are necessary for an understanding of why NASA's search has the strengths and weaknesses it does. If you, as an amateur, want to do it better, you must understand how to perform a sensitive search. Therefore, I will try to immerse you in the gripping details of signal processing without getting you lost.

## TAMING FREQUENCY AND TIME

What frequency does Middle C have? Is it the same for a piano, violin, and flute? If so, why do these instruments sound different? What frequency do you create when you whistle Middle C? What frequency do you produce when you snap your fingers or turn on a light switch? What is the frequency of a gunshot?

If you are certain of all the answers to the above questions, you probably understand the concept of frequency. Most people intuitively grasp what frequency is, since your ear is a frequency analyzer. To minimize the use of mathematics and to maximize the application of your intuitive understanding, I will use auditory examples in explaining what frequency really is. Of course, the concepts are equally valid when they are applied to fluctuations in radio receiver output rather than to those impinging on your eardrum as air pressure changes.

Many people know that Middle C corresponds to a frequency of 256 cycles per second. In other words, a Middle C note causes air pressure fluctuations that have 256 maxima and minima per second. This correspondence is arbitrary and by tradition. Your ear, with a precision varying widely among individuals, can tell the difference between Middle C and nearby notes. Some people even have what is called perfect pitch. Without any reference, they can tell you what note is being played on an instrument.

If you, like me, are not one of the gifted few with perfect frequency discrimination, how can you determine the frequency of a note? Usually this is done either by comparison or by direct counting of the wave maxima. If you have ever tuned a musical instrument, you probably understand the comparison method. As you tune your instrument more closely to a standard pitch, you hear a "beat" or difference in tone between the standard and the instrument being tuned. As the difference becomes small, it sounds like one note wavering in loudness at the frequency of the difference. You tune until the wavering stops. But what does "stop" mean?

For me, it means a wavering of less than once in ten seconds. However, to be sure that I have tuned to this precision, I must wait for ten seconds to make sure the note of the two sources is not wavering too fast. If I were willing to spend the time, I could wait a hundred seconds and get a more precise tuning. The point is that the precision with which a frequency is known depends on the time you have to measure it.

Counting the number of maxima gives the same result. If you wait a second, you can, by counting, know that your note produces 256 crests in that time, not 255 or 257. Thus, you know the frequency to 1 cycle per second (CPS). If you wait one hundred seconds, you can know the frequency to 1 cycle per one hundred seconds or 0.01 cycles per second. Again, the precision of the measurement is greater as the time span of the measurement increases.

This is a general principle of frequency measurement with applications in everything from radio to quantum mechanics. Whatever instrumentation you use, the product of the frequency error in CPS and the measurement time in seconds is always greater than or equal to one.

There is another subtlety to the concept of frequency. Commonly, people say that a piano and a violin playing Middle C produce the same tone. But everyone also knows that these two instruments have characteristic sounds that allow a musician and most other people to tell instantly what instrument is being played. The difference between the notes lies in the overtones of the sound. Both instruments produce sine waves at 256 cycles per second when playing Middle C. However, they also produce weaker sine waves at multiples of this frequency, 512, 768, 1024 cycles per second, etc. It is the relative strengths of these overtones that give each musical instrument its characteristic sound.

A pure frequency, like the one you whistle, is a single sine wave. It is possible, in fact, to create any wave shape, no matter how complicated, by adding sine and cosine waves together at appropriate frequencies and with appropriate strengths. Nowadays, makers of electronic music apply this principle with abandon. By adding together the right combinations of sine waves, they can synthesize the sound of any acoustical musical instrument ever made. They can also synthesize sounds that no physical acoustical music box could ever make. For those of us who like this sort of thing, this is much of the appeal of electronic music.

So far we have unearthed two important principles of wave analysis. First, the frequency accuracy depends on the time interval over which it is measured. You cannot assign a frequency to snapping fingers or a gunshot because the sound is too short for accurate determination. Second, any wave shape can be synthesized as the sum of many simple sine waves with specific frequencies and amplitudes. This latter principle is very important in signal analysis because physical systems, whether in radio or music, tend to produce signals with large amounts of a restricted set of frequencies. A musical instrument, a TV station, or your local AM broadcast station all produce primarily one "note." If one analyzes these signals to determine what frequencies and amplitudes are present, their strong primary notes stand out as obviously artificial in a natural world filled with random noises covering a wide range of frequencies with no outstanding components.

Computers use a program called the FFT or Fast Fourier Transform to efficiently perform signal analysis in terms of sine waves. Fourier discovered the mathematical relation between arbitrarily shaped wave forms and sine waves in the early 1800s. Multiplying the output numbers from the transform by the appropriate sine or cosine wave and adding up all the results exactly reproduces the input data. In other words, the transform tells you how big each pure frequency is in a signal, compared with all the others. The computer always works with a finite time slice of the input. (We can't observe forever!) The output from that data, the amounts of sine and cosine waves present, is the frequency spectrum of the signal. Each number in the output applies to a particular frequency. The spacing of the frequencies in the analysis is consistent with the measurement principle for frequency. If the spectrum works on data that extends over one second, the spacing of the frequencies (the "bin size") is one cycle per second. If the data extends over 100 seconds, the spacing of the frequencies is one cycle per hundred seconds or 0.01 cycles per second.

We can imagine carving up a batch of data into time segments and running an FFT on each one. The data is then represented by the amount of sine wave amplitude for each frequency at a particular time. The sine wave amplitude is the amount of a particular frequency needed, during a particular time interval, to synthesize the data there. The output of the Fourier Transform can thus be viewed as an array of numbers with rows representing time and columns representing frequency. Each number is the sine wave amplitude in its own little frequency-time rectangle. The only restriction on the rectangular areas is that they satisfy the frequency precision rule. Fat or thin, the frequency interval multiplied by the length of time for the measurement must equal one.

This procedure would not accomplish much except that artificial signals of many classes look very different from the natural noise of the Universe when analyzed in this way. Noise in general is broad banded, with many frequencies present in about equal amplitude. If you tune your FM radio to a place on the dial where there is no station, you will find a roar like incoming surf or a waterfall. It sounds nothing like a whistle or note on a violin. This is because no particular frequency dominates. In fact, all frequencies in the audible range for the human ear are present equally. Thus, in analogy with white light, in which all colors are equally present, the static in your FM receiver is called white noise. This noise is characteristic of most natural objects. Performing a frequency analysis is not effective on pure noise.

A signal behaves quite differently, however. In the input, the signal, even if it is a pure sine wave, is spread out over time. But in the output, only one or at most a few frequencies contain all the signal amplitude. Thus, a few bins contain all the signal. The noise in these same boxes, however, is only a small fraction of the total in all the data analyzed, since this noise is distributed equally throughout the data stream. Thus, at frequencies where the signal is present, the SNR of the output is much larger than at the input to the Fourier Transform.

Not all signals can be concentrated by a frequency analysis. FM signals, by their very nature, defy this type of processing. FM means frequency modulated. The output of an FM transmitter is a carrier whose frequency changes in proportion to the amplitude of the modulating signal. Thus, the transmission is spread over a wide band of frequencies. No frequency is occupied long enough to achieve much amplitude. Observing the band with an amateur scanner can be more sensitive than any NASA narrow-band detector proposed for the 1990s. On the other hand, AM radio and TV have a single sinusoidal component (the carrier wave) containing over half the power of the transmission. Though this unchanging continuous wave part of the transmission has no information content, it makes the construction of cheap receivers possible.

If carriers are commonly used by technological civilizations, they are like no natural sources known. They can be made very pure, covering less than .001 cycles per second. A long Fourier Transform of 1,000 seconds produces frequency channels well matched to such transmissions. Any carrier is enormously large in its channel, and the noise is small.

There is a problem with looking for stable ET carriers in this way. First, it is not always possible to look at sources for 1,000 seconds. Second, frequencies as we observe them are not the same as the transmitted frequency. The ET, especially if transmitting from an Earth-like planet, is in relative motion with respect to our solar system. We can correct for the motions of the Earth relative to the center of our solar system, so this is no problem. However, we do not know the velocity of the ET source. As this velocity changes with the orbit of the ET world, the apparent frequency we see changes. This frequency shift is due to the Doppler effect. It is the same effect that causes a car horn to change pitch as a vehicle passes you or a whistle of a train to drop in pitch as it passes along its track. Thus, if each channel is too narrow, a signal will not remain there during the analysis and will drift over many frequencies instead of staying in just one.

The NASA search has compromised. For nearby Sun-like stars that are observed for a total of many minutes, the bandwidth of the frequency bin is about one cycle per second. This means that each second a new spectral analysis is made. Within one second, even a rapidly rotating or orbiting planet will not change velocity enough to move the received signal out of the 1 cycle per second bin. For stars that are to be scanned very rapidly, in the all-sky survey, the frequency channel is 32 cycles per second wide. This means that a spectrum is computed every 1/32 second.

Let's take a particular example that could represent a possible ET signal. Assume that the closest communicating civilization is a few hundred light-years away, sending a pretty strong signal. Since it is not one of our thousand nearest Sun-like stars, it is not in any catalogue. Thus, we will not do a high-sensitivity search in its direction using a large antenna and looking for a long time. It will be left, instead, for the attention of the all-sky survey.

This survey uses a moderate-sized antenna, thirty-four meters, about ten times the diameter of a typical TV satellite dish at three meters. The folks at JPL, part of the NASA SETI program, use this smaller antenna to survey the whole sky at moderate sensitivity.

The hundredfold advantage in antenna area for the thirty-four-meter dish means that JPL and an amateur looking at the same star would see a hundredfold difference in signal strength, in favor, naturally, of the professional. However, in order to cover lots of frequency, JPL made each of its frequency channels 32 cycles per second wide. This means that the 10 million channels in their machine could cover a band 300 million cycles per second wide. This allows coverage of their desired 9 billion cycle per second frequency range, by moving the analyzer in thirty nonoverlapping frequency steps.

## INDIVIDUAL AMATEUR SEARCHES

Because they must cover the whole sky over a wide frequency range, the NASA all-sky survey moves the analysis band and direction rapidly, spending perhaps 1 second at each directional setting. If you were certain about your star and channel, you could actually equal or surpass their sensitivity for that frequency and direction by spending more time and processing power there.

In particular, if you knew the frequency of the ET signal exactly, or even approximately, you could pour data from the output of a commerically available receiver such as an Icom ICR7000 into a digitizer attached to your PC. The approximately 3,000 cycle per second audio bandwidth could be sent through a computer program that synthesizes thousands of little filters. Each filter could be 0.32 cycles wide, one hundredth the width of a JPL bin. The output signal to noise ratio in the correct channel would be just as good as that for the pros at JPL. The reason is that while they have a bigger antenna and one hundred times the signal, you have one hundredth the filter width and one hundredth the noise power. You can "hear" as well as any search is likely to do in our lifetime. The cost of this improvement is the narrower width of the FFT analysis channels and longer observation time. JPL gets thirty-two samples in their one-second look at a particular star and frequency. Your bandwidth is one hundred times smaller so each analysis takes one hundred times longer. Remember the accuracy rule. Thus, to get the same number of spectra, thirty-two, it will take you one hundred seconds. The reason each of your channels has the same energy that the pros have is that your analysis allows a hundred times as long for the signal to put energy into the channel. Thus, though the received signal energy per second is small because of your small antenna, it is compensated for by the longer time. You also probably own your antenna *because* you live in a rural area, which not only has no commercial TV but also no human interference—an advantage for the search!

Of course, the problem with all this is knowing the exact frequency of the signal to camp out on. If the signal is drifting because of the motion-induced Doppler effect, your detection attempt will fail (unless you have a drift algorithm to match it) since the frequency will change while the analysis is being carried out. On the other hand, an extraterrestrial might intentionally stabilize his signal. He could either take out the Doppler shifts in the direction he is transmitting, just as we do for receiving, or he could transmit from a stable platform in deep space.

### BEAMED BEACONS

We know how strong a signal must be if we are to see it from Arecibo. The calculation is somewhat complicated, taking into account everything from antenna temperature to number of samples added. The net result is easy to state, however. The power of an ET transmitter broadcasting in all directions at once, just visible from Arecibo with NASA's SETI analyzers, is

$$P = 10R^2$$

where $R$ is the distance of the transmitter in light-years and $P$ is in megawatts. The observation time is one hundred seconds. For a star twenty light-years away, (there are about twenty Sun-like stars that close), the required power is four billion watts. This is the total output of two large nuclear power plants; possible, but not trivial either.

This, however, is the answer if the power is distributed by the sender uniformly over the whole sky. It goes out in all directions from the alien transmitter and just enough falls on the little portion of that enormous surface intercepted by the Arecibo antenna. There is just enough signal for detection. What if, instead, the alien civilization *beamed directly at Earth,* either because they had already received one of our own radio signals or because we orbit a Sun-like star. If they used an antenna like Arecibo, concentrating their energy a millionfold, the required transmitter power would be only four thousand watts. This is a power level many radio amateurs can produce with their rigs at home now, though in the U.S. it is illegal to use this much. There could be an intermediate case: a transmitter placed in the galactic plane, broadcasting an annular beam pattern that just irradiates the disk, might cut power requirements by factors of five to twenty-five times. Even a three-meter dish could receive a four-billion-watt signal from their Arecibo equivalent.

Though such levels are not impossible, they may not be common. Obviously, if civilizations are to be seen at greater distances, the power levels must be even higher or observation times must be longer. The latter is where amateurs can shine.

The important thing to remember is that most frequencies have not been searched even at the amateur level I am suggesting here. Most people using these frequencies are transmitting signals and looking for particular, rather strong, return signals. You can conduct, on most frequencies up to 1.3 billion cycles per second, a search thousands of times more sensitive than any done before, and you can do it with commercially available amateur equipment. If there is a stable, beamed signal out there in the solar neighborhood, you could be the first to find it.

Before you get carried away and jump in your car for a trip to Radio Shack, remember that this only applies if you know the star and the frequency. In thirty seconds, JPL will be able to survey the particular point in the sky where you are looking, over a frequency band of ten billion cycles per second. To survey the same band with your amateur set up at the same sensitivity, you would require ten years. Do you see why an amateur effort is like the lottery? You could spend your lifetime playing and never win.

## AMATEUR ANALYSIS OF THE NASA DATA STREAM

There is another way that amateurs, even the public at large, might potentially participate in SETI. As mentioned, NASA has devised algorithms to analyze incoming data in real time (largely the work of the SETI Signal Detection Team). These detection methods filter out false alarms due to noise and terrestrial interference and recognize a simple signal. Experimentally, these techniques have been shown to be many millions of times faster for the low false alarm rates required than human searches are. Such a human search, for example, might look at the data on a video display as it comes in. Generally, any signal that is not a continuous carrier wave or a set of pulses will not be sensitively detected by the NASA algorithms. On the other hand, the human eye is very good at finding unspecified nonrandom patterns in noise. Just as a longshot then, it might be interesting for

NASA to broadcast (over "SETI-TV") some of the SETI data. If you were looking and thought you saw a pattern, you could note the time and phone in a report to NASA or an amateur/professional coordinating group at an 800 number. Judging by the reliability of UFO reports over the years, this idea has obvious problems, but amateurs with an original computer idea could tape this data for off-line (not real time) analysis. Anyone with a PC and the right equipment could try some tricks on selected data ... if NASA chose to broadcast it.

NASA has no plans to routinely use human observers, because, though the human processor is very creative in identifying patterns, the flip side of the coin is a high false alarm rate. Humans tend to find patterns, even when they are not there. Nonetheless, trained operators will sometimes look at data directly, even in the automated systems. In the end, we never believe data not accessible to our senses. NASA currently has no plans for SETI-TV or an 800 number. This would require funding and coordination that do not currently exist.

There are two further caveats about this whole idea. Since a pulse or carrier wave is so simple to detect, it seems the likely choice for any ET beacon. Thus, any complex signal sent out is likely to be leakage, not a beacon, and thus weak to begin with. In addition, the modulation that carries the complex information will be weaker than the carrier, if there is one (which NASA is likely to see first). So recognizing a "picture" seems unlikely. On the other hand, the suggestion about special reprocessing of the data may be feasible.

## STEPS TO AN AMATEUR SETI OBSERVATORY

If, after the dose of cold water in the section on beaming, you still want to build a system, the parts are readily available. The receiver, an ICR7000, is commercially available. It costs $1,000. The computer is a PC compatible, another $1,000. The digitizer to get your data into the computer is $100. The amplifier to get a good effective (low) noise temperature is $100. The mixer, if you want frequencies above a billion cycles per second, is $100. The satellite dish is about $500 (maybe you already own one). So for $3,000, you're off and running. This includes the cost of the filter bank synthesis program called a DFT. For a three-meter dish and thirty-two seconds integration time, the sensitivity is better than $10^{-22}$ W/m$^2$. This yields a few false alarms per hour using the voice channel of the ICR-7000.

## DIRECTIONAL CONTROL

Pointing your antenna is easy. Anything that points a three-inch optical telescope is more than accurate enough for a radio telescope. (Mechanical strength is, however, another issue.) Radio waves are much longer than light waves. Even though your radio dish is tens of times bigger than a three-inch refractor, the waves it is focusing are millions of times larger than light waves. Therefore, the focus is much less precise for the radio telescope. You do, however, want good time signals and pointing coordinates, so that you and others can look again in a direction where you found a signal. Setting circles, or measured displacements from known celestial radio sources, will suffice (see Appendix).

## TRACKING THE SKY

Amateur searches for ETI signals can be carried out in various ways. Most TV dishes are permanently pointed toward a satellite. They can be used for SETI as stationary "transit instruments" that detect sky sources crossing the antenna beam as the Earth turns. This mechanical simplicity comes at a price. The satellite, which is interference for SETI purposes, is smack in the middle of the beam. This makes frequencies near its transmission frequency unobservable. Second, in the transit mode, each object will be in the beam for only a few minutes per day. Furthermore, if the dish is not repointed in declination (celestial altitude, with respect to the celestial poles), only one strip of the sky will be surveyed.

In order to attain more sensitivity, the radio dish should track the celestial source under study. To do this, a mechanical drive is needed. Pointing a radio antenna is easier than pointing an optical one because of the broader radio beam and because atmospheric effects are negligible for radio waves. However, as with optical tracking, the drive has to run one revolution per sidereal day, slightly shorter than the solar day.

Accurate setting circles would also be useful. If these are not available, a PC with the right program can feed pointing data based on looking first at a celestial source with known coordinates.

We can give no single recipe for adapting the dish to a drive—there are too many dish types on the market. It may be that one of the new types of "table drives" can be adapted to support a whole dish without disassembling its mount, thus preventing the invalidation of equipment warrantees. (See Appendix.) Perhaps further work on such drives can be coordinated through the ATM column of *Sky and Telescope* magazine, or through the SETI Institute or another coordinating group.

## SETTING UP AMATEUR GANGED SEARCHES: BIG COOPERATION

In spite of the simplicity and potential glory that might point you to individual searches, you should consider cooperative efforts. These can yield benefits in better sensitivity (signal to noise), time coverage, sky coverage, frequency coverage, and even spatial resolution, though the latter is probably less important and is more difficult.

## TRACKING VERSUS TRANSIT

As mentioned, if a selected set of targets can be agreed upon, you and others can track and "camp out" on them. Requiring many simultaneous detections of the same signal will allow each individual detector to use a lower threshold without increasing the overall false alarm probability. This increases total system sensitivity. If standard time marks are imposed and standard equipment and algorithms are used, then, in principle, tapes of signal intensity from many dishes can be combined to give greater signal strength. (Combining phase information would yield even greater gains. Unfortunately, this is presently beyond the state of the amateur art, but who can say what time and frequency standards may be cheaply available in ten years?) Even if signals are not phased together, the effective SNR improves in proportion to the square root of the total dish area observing simultaneously. If the dishes are all the same (say three meters diameter), then the improvement goes as the square root of the number of participants; ten participants working simultaneously on one target will have an SNR about three times better than one observer. Given a nondrifting constant signal or one with long repetition period, daily reobservations can be added together to further improve the SNR.

Further cooperation could search the same direction, parceling out various search frequencies to allow, for example, the tracking of drifting signals. Alternatively, round-the-clock observation would be possible with an observation network spanning the globe. Observers over a range of latitudes can cover the whole sky in both hemispheres.

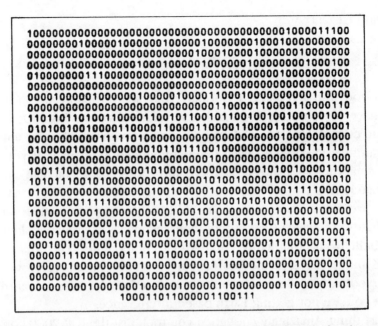

```
100000000000000000000000000000000000000000001000011100
00000000100000100000010000010000000010001000000000
00000000000000000000000000001000100001000001000000
00000100000000000100010000010000001000000001000100
01000000011100000000000000100000000000010000000000
00000000000000000000000000000000000000000000000000
00001000001000000100001000011000100000000110000
0000000000000000000000000001100001100001100000110
1101101101001100011001011001011001001001001001001
0101001001000011000011000011000011000011000000001
00000000001111101000000000000000000010000000000
01000001000000000001011011100100000000000001111101
0000000000000000000000000001000000000000000001000
1001110000000000001010000000000000001010010000100
1010111001010000000000010100100001000000000010
01000000000000000100100000100000000001111100000
00000000111110000001110101000001010100000000010
1010000000100000000000010010000000101000100000
0000000000001000100100010001001101100111011011010
0000100010001010101001000100010000000000000010001
0010010010001000100000010000000000011110000011111
0000011100000001111101000001010100000101000010001
0000010000000000100000100001110001000001000001100
00000000100001000100010001000001000011000110001
000001000100100010000010000011000000001100001101
1000110110000011100111
```

Figure 1-A. An example of a digitally encoded message. It has the look of randomness.

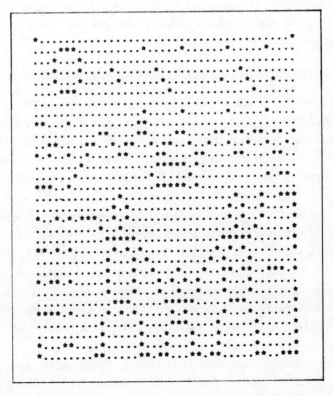

Figure 1-B. The decoded message of 1-A. Method used was to find repeating information, which first suggested arrangement of the data as a TV frame. Through further decoding, a picture emerged. *(Generated by Cipher A. Deavour.)*

## OTHER DATA

Could you find ET using a normal radio set? Probably not. After all, people have been tuning the bands for years. If signals from Out There were loud enough for normal electronics to find without directive antennas and computers, they would have been announced long since. If you don't want a construction project, *is* there any other data you can analyze? Unfortunately not.

If you are rich, a system such as SERENDIP complete with frequency shifting hardware, which surveys fifty thousand cycles per second bandwidth, can be built from completely commercial components. The total cost is less than thirty thousand dollars (exclusive of the antenna). However, I refuse to look at your low-probability events unless you take extraordinary care that you are not getting Earth-based interference.

One last thing. Amazingly enough, if you took just the ICR-7000 commercial receiver and listened for strange things on its output with a satellite antenna in standard scanning mode, you would be surveying most frequencies and positions in the sky at one thousand times the sensitivity anyone has ever used before. There could be a strong ET out there somewhere, just waiting for the right dish to point at the sky and listen to the right frequency. Maybe you will get lucky.

## COORDINATION OF AMATEUR SETI WORK

We have suggested possible ways amateurs can work on SETI as individuals or in groups. In either case, coordination will be needed to help guarantee high-quality observations and, for ganged searches, a high level of standardization. No organization currently exists to do this. Amateurs should be gentle in approaching professionals about this, as they already have their hands full.

Analogies for such an effort do exist. The Audubon Society coordinates amateur bird observations and is a resource for professionals in that field. The Amateur Association of Variable Star Observers (AAVSO) has existed for many years, making important contributions to professional astronomical observing programs using standard techniques. Also, amateurs regularly search for new comets and supernovae. Their occasional discoveries are reported over the International Astronomical Union (IAU) telegram alert network.

Who could coordinate amateur SETI efforts? There are several possibilities. Among existing groups, the IAU Commision 51 on SETI, the SETI

Institute, the Astronomical Society of the Pacific, and the Planetary Society are possibilities. The last may be the best, as it has broad aims, with strong amateur and professional participation. Its executive director, Thomas McDonough, has expressed preliminary interest. He is not equipped to deal with massive interest, however. There are 2.4 million satellite dishes out there in the United States. If only 10 percent of those dish-owners decided to join in a ganged search, the combined effort would produce the equivalent of 25 Arecibos—but the job of coordination would swamp the Planetary Society. The simplest thing to do now is to become a dues-paying member of one of the organizations listed above and enclose a written expression of interest in working on SETI—along with any suggestions.

## WRITE YOUR CONGRESSIONAL REPRESENTATIVE

For those of you who have given up on doing SETI yourself by this time, realizing just how daunting the search size is, please write your congressional representative if you feel that an ETI search is worthwhile. A few letters have real impact in Washington, especially if they are carefully thought out. With luck, the big NASA search will have begun by late 1992, Congress willing. A letter from you might make the difference.

# CHAPTER 10

# THE NEED
# TO KNOW

*The modern era of SETI began with the classic paper of Giuseppi Cocconi and Philip Morrison in 1959. Professor Morrison gives us the benefit of his thinking on the subject here, preceded by Arthur C. Clarke, who has probably done more to make the public aware of SETI than any writer of any generation.*

*The concluding words of Cocconi and Morrison's seminal paper are as valid today as they were when first written, more than three decades ago:*

> *Few will deny the profound importance, practical and philo-sophical, which the detection of interstellar communications would have. We therefore feel that a discriminating search for signals deserves a considerable effort. The probability of suc-cess is difficult to estimate, but if we never search the chance of success is zero.*

# WHERE ARE THEY?

## BY ARTHUR C CLARKE

Early in December 1985, a group of distinguished astronomers gathered in Colombo, under the auspices of the International Astronomical Union, the Institute of Fundamental Studies, and the Arthur Clarke Centre, to discuss a subject which has long fascinated the general public, but which has only become scientifically respectable during the last two decades. I refer, of course, to the possibility of life on other worlds.

Now, this is a fairly new idea in Western thought, for the simple reason that from Aristotle onward it was assumed that Earth was the center of the Universe and that anything beyond it was some vague celestial realm inhabited only by supernatural beings.

The Sun was obviously a mass of fire, so no one except the gods could live *there*. And as for the moon, it probably wasn't big enough for many occupants. . . .

The five planets visible to the naked eye—Mercury, Venus, Mars, Jupiter, Saturn—had been known to mankind since prehistoric times, and their curious movements had been a cause of much speculation. But no one, except a few eccentric philosophers, had any idea that they were all worlds in their own right—two of them enormously larger than the terrestrial globe.

It's an extraordinary fact that the East had guessed the true scale of the Universe, both in time and in space, centuries before the West. In Hindu philosophy there are eons and ages long enough to satisfy any modern cosmologist; yet until only a dozen generations ago much of Europe believed that the world was created around 4000 B.C. (I'm sorry to say that, owing to their misreading of the Bible, thousands of foolish people still believe such nonsense.)

The turning point in our understanding of the Universe may be conveniently dated at 1600, just before Galileo pointed his first telescope toward the stars. Shakespeare belongs to the century before the great intellectual revolution.

> *Doubt that the stars are fire,*
> *Doubt that the sun doth move . . .*

he wrote, circa 1600. He was wrong on both counts.

Of course, we know that the Sun *does* move—but not in the way that Shakespeare imagined. He thought it moved around the Earth, as common sense seems to indicate, and had no idea how distant—and how big—it really is.

And the stars aren't fire—although for reasons that were not understood until well into this century. They are much too hot! Fire is a *low*-temperature phenomenon in the thermal range of the Universe, much of which is simmering briskly at several million degrees, where no chemical compounds can possibly exist.

During the seventeenth century, the telescope revealed for anyone who had eyes to see that the moon provided at least one other example of a world with mountains and plains, though not rivers and oceans. The moving points of light that were the planets now turned out to have appreciable discs—and one of them, Jupiter, had its own retinue of moons. Clearly, Earth was not unique; nor, perhaps, was the human race.

This was a shocking—even an heretical—thought, at least to those brought up in the Aristotelian school. Anyone who preached it too loudly, and especially near Rome, was likely to get into serious trouble with the Inquisition.

The classic example is, of course, Giordano Bruno (1548–1600), who was one of the first European advocates of the doctrine of an infinite Universe and the "Plurality of Worlds." Refusing to recant, he was burned at the stake in 1600. I wonder how many modern scientists would be prepared to emulate him in defense of their theories. . . .

And while we're on the subject of Renaissance astronomers, I'd like to remind you one of the greatest ironies in the history of science.

In 1582 that remarkable man, Father Matteo Ricci, arrived in China with all the latest wisdom of the West. The Chinese regarded Occidentals as barbarians (and probably still do, though they're too polite to say so). But by tact, intelligence, and sheer goodness Father Ricci persuaded them that their superstitious concept of an enormous Universe enduring for vast eons was all nonsense: God put the Earth in the center of everything, and Adam and Eve in the Garden, only a few thousand years ago. As he proudly wrote: "The Fathers gave such clear and lucid explanation on all these matters which were so new to the Chinese, that many were unable to deny the truth of all that they said; and, for this reason, the information on this matter quickly spread among all the scholars of China."

Well! Poor Father Ricci! While he was persuading the Chinese to take a Great Leap Backward to Ptolemaic astronomy, Copernicus was destroying its very foundations in Europe. A few decades later Galileo (with an anxious glance over his shoulder at Bruno) would finish the job of demolition.

For the last three hundred years, not very long in human history, all educated persons have known that our planet is not the only world in the Universe, and that its Sun is one of billions. The great voyages of the seventeenth and eighteenth centuries, during which European explorers "discovered" whole cultures that didn't even know they'd been lost, also prompted speculation about life on other planets. It seemed only reasonable that our enormous cosmos must be populated with other creatures, some of them perhaps far more advanced than we are. The alternative—that we are utterly alone in the Universe—seemed both depressing and wildly megalomaniac.

But how to prove it, one way or the other? We children of the Space Age can no longer remember how enormous even the solar system seemed, only a lifetime ago. Now the Voyager space probe is heading for its appointment with Neptune, which, as recently as 1930, marked the frontier of the Empire of the Sun. That is an impressive achievement; even so, it will be tens of thousands of years before Voyager can cross the gulf to the nearest star.

Fortunately, we do not have to rely on *physical* contact to discover if there is intelligent life elsewhere in the Universe. We now assume, almost as a matter of fact, that any contact is likely to be by radio. Yet this in itself would have seemed incredible until well into this century. We take radio so much for granted that we forget how miraculous it is. Even the most farsighted prophet could not have predicted it—which is yet another example of what I call Comte's Fallacy.

Around 1840 the French philosopher Auguste Comte (1798–1857) was rash enough to make the following pronouncement about the limits of our knowledge concerning the heavenly bodies: "We see how we may determine their forms, their distances, their bulk, their motions, but we can never know anything of their chemical or mineralogical structure; and much less, that of organised beings living on their surface . . ."

Comte's monumental gaffe was in the same class as Father Ricci's. Within a few decades, the invention of the spectroscope had utterly refuted his assertion that it was impossible to discover the chemical nature of heavenly bodies. By the end of the century, precisely *that* was the main occupation of most professional astronomers. Only the amateurs were still concerned with what Comte believed must always be the entire body of their science.

So it is very dangerous to set limits to knowledge or to engineering achievements. No one could have anticipated the spectroscope; and no one could have imagined radio. They both exemplify Clarke's well-known Third Law: "Any sufficiently advanced technology is indistinguishable from magic."

There may be "magical" inventions or discoveries in the future which will settle the question of intelligent life in the Universe, but I do not think we really need them. Today's electronics can probably do the job, given a few more decades of determined application.

The giant radio telescopes which have been built for purely scientific purposes are quite capable—and this is a splendid example of serendipity—of detecting the sort of radio signals one would expect from an advanced civilization in our immediate galactic neighborhood. It would be ridiculously optimistic to expect immediate success, since we have had the capacity of making such a search for less that half a human lifetime.

Yet already this "failure" to find an artificial signal has produced a kind of backlash and has prompted some scientists to argue, "Perhaps we *are* alone in the Universe." Dr. Frank Tipler, the best-known exponent of this view, has given one of his papers that provocative title "There Are No Intelligent Extra-Terrestrials." Dr. Carl Sagan and his school argue (and I agree with them) that it is much too early to jump to such far-reaching conclusions.

Meanwhile the controversy rages; as has been well said, *either* answer will be awe-inspiring. The question can only be settled by evidence, not by any amount of logic, however plausible. I would like to see the whole debate given a decade or two of benign neglect, while the radio astronomers, like gold miners panning for dust, quietly sieve through the torrents of noise pouring down from the sky.

There is also another, and much more speculative, line of approach to this problem. Let me give an analogy to explain what I mean.

If a visiting traveler had surveyed our planet from space ten thousand years ago, he would have seen many signs of life—forests, grasslands, great herds of animals—but no trace of intelligence. Today, even a casual glance would reveal cities, roads, airfields, irrigation systems—and, at night, vast constellations of artificial light.

These "advertisements" of terrestrial civilisation would have been beyond the imagination of our Stone Age ancestors. Can one set any limits to what might be achieved by a really advanced, long-lived society, with thousands of centuries of space-faring behind it? In particular, might it not have—literally—set its sign among the stars, as we have done upon the Earth? As long ago as 1929 the physicist J. D. Bernal, in one of the most daring works of scientific imagination ever penned, wrote: "It is unlikely that man will stop until he has roamed over and colonised most of the sidereal universe, or that even this will be the end. Man will not ultimately be content to be parasitic on the stars but will invade them and organise them for his own purpose . . . By intelligent organisation the life of the universe could probably be prolonged many millions of millions of times. . . ."

Later writers have talked about "the greening of the Galaxy," and asked why the stellar sky is so untidy and badly organised. Where, indeed, are the Cosmic Engineers?

Perhaps, like ants crawling around the base of the Empire State Building, we simply haven't recognized what's going on all about us. During the last few decades, astronomers have been discovering some very strange phenomena in space, and have been straining scientific theories to the limits in attempts to find *natural* explanations.

Pulsars were the first example. In 1967, when Hewish and Bell discovered radio sources ticking away more accurately than any mechanical clock ever made by man, their first wild speculation was that they might be artificial. Indeed, no astronomer before 1920 could have explained how unaided Mother Nature could have contrived such a prodigy.

Well, we are now quite sure that pulsars are indeed natural (though amazing) objects—tiny dying stars acting like cosmic beacons as they spin madly on their axes. But there are other phenomena not so readily explained, and I should not be in the least surprised if the astronomers finally give up on them and admit: "We're sure that Nature can't be responsible. Somebody out there has forgotten to switch off the lights."

Or worse. The most chilling explanation I have heard of one titanic outpouring of cosmic energies is: an industrial accident. . . .

Nowadays, anyone who considers that alien supercivilizations may exist has to contend not with skepticism but with something much worse—credulity. Although the subject now affects me with uncontrollable fits of yawning, I would be failing in my duty if I did not say something about UFOs.

So here, as briefly as possible, are the conclusions I've come to after more than fifty years of study. (Fifty-six, to be exact, since I first read Charles Fort's *Lo!* in 1930. That monument of eccentric scholarship, published long before anyone had ever heard of flying saucers, listed apparent celestial visitations right back to the Middle Ages.)

1. There may be strange and surprising meteorological, electrical, or astronomical phenomena still unknown to science, which may account for the very few UFOs that are both genuine and unexplained.

2. There is no hard evidence that Earth has *ever* been visited from space.

3. If that *does* happen, there are at least three independent global radar networks that will know within a matter of minutes. And in the unlikely event that the U.S., U.S.S.R., and Chinese authorities instantly cooperate to suppress the news, they'll succeed for a maximum of forty-eight hours. How long do you imagine such a secret could be kept? Remember how quickly Watergate unravelled.

Having written thousands of words on the subject (and read millions) I refuse to go into further details. If anybody wants to argue, I'll merely quote one of my favorite book titles: "Shut up, he explained."

Finally, if *they* are out there—what do they look like? I suggest you go to the local zoo and take your choice. Nature tries everything at least once—and has lots of time and space for experimenting.

But I will tell you what they will *not* look like. We now understand the principles, if not the details, of human evolution. We specimens of *H. sapiens* are the product of thousands of successive throws of the genetic dice, any one of which might have turned out differently. If the terrestrial experiment started all over again at Time Zero, there might still be intelligence on this planet, but it wouldn't look like us. In the dance of the DNA spirals, the same partners would never meet again. As Loren Eisley wrote in *The Immense Journey* over thirty years ago: "Nowhere in all space or on a thousand worlds will there be men to share our loneliness. There may be wisdom; there may be power; somewhere across space great instruments . . . may stare vainly at our floating cloud wrack, their owners yearning as we yearn. Nevertheless, in the nature of life and in the principles of evolution we have had our answer. Of men elsewhere, and beyond, there will be none forever. . . ."

# REFLECTIONS ON THE BIGGER PICTURE

## BY PHILIP MORRISON

One of the most important occurrences over the fifteen or twenty years during which we have been thinking about SETI has been the event that did not happen: the discovery of some form of life on another planet. What has occurred that we did not foresee, but that supports our general view while complicating our detailed quantitative understanding, is the finding of complex polyatomic carbon compounds in a wide variety of cosmic contexts. From studies of the molecular species in the giant dense galactic clouds and from analysis of some asteroids and maybe satellites (in which we find indications of carbonaceous compounds), we have derived a renewed interest in carbonaceous meteorites themselves. This demonstrates a point on which biochemists were already very clear fifty years ago: The flexibility and peculiar subtlety of carbon compounds, as well as the high abundance of carbon, makes them preeminently the source of complex chemistry in the Universe.

When we seek the input of the physical sciences, we know that the practitioners are masters of a powerful deductive structure, with quantitative possibilities. Of course, they are beset by the necessary complexity of their models. They must try to pin down from some a priori model just what the first six or seven hundred million years of Earth history were like. They search for a necessary prelude for the biologists, something essential to the total picture of cosmic evolution. On the other hand, if the biologists were to explain why they needed certain conditions, perhaps it could be determined whether these conditions were at all plausible under some existing model. It is clear to me that strong interchange must go on in this domain. We have to learn a more interdisciplinary way of facing such problems.

A famous nineteenth-century interdisciplinary dispute illustrates this point. There was a lot of dispute, even full conflict, but no resolution at all; logic was clearly on one side, yet that side turned out to be wrong. I am talking about the famous problem of the time scale available for Darwinian evolution.

The absence of substantial observed changes in speciation in the natural world, as compared with the swift changes produced by domestic hybridization during the course of history, is a strong argument for the slowness of natural speciation. This argument was made even stronger by the fact that paleontology showed change clearly; therefore, you had to say that the time available in the geological record was very large. Darwin, while he was a man of extraordinary logical ability and in very simple ways a brilliant experimenter, often cutting right to th heart of the matter, was a bad mathematician. He could not calculate anything. He had a touching faith, however, in instrumental methods. His son writes that he discovered his father making measurements with an old paper ruler and writing these down to high accuracy. The son commented. "Well, you know that ruler is probably not right." So he got a better ruler; sure enough, his paper ruler was stretched and deficient. Darwin was depressed; the notion that a calibrated ruler, a thing you trust to measure with, might not be right was a breach of faith that he could hardly accept from his ruler-making colleagues! That was his style. He kept saying that his study of geological records suggested that there was a great deal of time indeed. He liked to put it that biology required almost infinite time. By this he meant a time quite long compared with all times that had been suggested so far.

Now comes Lord Kelvin, armed with the most powerful physics of the nineteenth century, who was able to show with computational hammer blows, one after another, that the time available for Darwin's evolution could not be one hundred million years, it could hardly be sixty million years, perhaps not even ten million years. For, if the Sun burns carbonaceous fuel or any similar chemical fuel, it can only last thousands of years—a palpably inadequate stretch of time. Even if it derives its energy from gravitation, we know its mass and we know its size, and we can show that it cannot last for more than a few tens of millions of years. Thus there was a direct conflict. Kelvin would come to the biologists and heckle them terribly by saying, "Tell me what time you want and I'll show if it can be; I can calculate everything." (He meant cooling and so on.) Of course, they couldn't name a time. They simply said, "Well, we know you're wrong. We feel it in our bones, in our fossils, but we can't prove it." So the evolutionists were regarded as people without any quantitative basis for their science, though they were, of course, onto something profound.

That was noticed by the distinguished geologist T. C. Chamberlain immediately after the discovery of radioactivity, about 1900. By 1905 or so, Ernest Rutherford himself gave a famous evening talk. It was a most distinguished and formal lecture to the Royal Society of London. As a young breaker of rules, a young discoverer, he noticed that in the front row sat Lord Kelvin himself, very elderly, but very stern, trying hard to stay awake and check on this young man who was going to talk about new forms of energy. (Kelvin didn't like radioactivity either, by the way.)

Rutherford tells us that he wondered how to avoid offending Lord Kelvin. Perhaps the famous man would get up and leave when Rutherford talked about the fact that nuclear energy can keep the Sun going for a good long time. Finally Rutherford thought of the right formulation. He said that he had been able to solve an old problem, whose magnitude and importance had been shown by Lord Kelvin when he pointed out that there was no known source of energy capable of keeping the Sun going. At last by experiment they had been able to find a new source, bringing out fully what Kelvin had shown all those years ago. Says Rutherford, Kelvin immediately went to sleep and the whole session was a huge success.

Let us look at a strategy: the strategy for catching food as a predator does. We are, after all, predators, both on berries and on bears; we belong to the hunting-foraging creatures. Suppose you were in the position of living on lobsters. That is a nice position to be in: It is achieved by some New Englanders and by all common octopi. Octopi are intelligent invertebrates—in some ways our analogs within the invertebrate kingdom.

Let us approach the octopus from the standpoint of a rational analysis of prey-seeking behavior. What is the situation? Lobsters are not as variable in behavior perhaps as some land animals, but still they are not all that uniform, either. It is quite likely that any particular desirable game, like lobsters, appears in fluctuating numbers within the field of action of any carnivore. There is little likelihood of a steady flow of lobsters, one dropping down every hour. Most hunters don't find it that easy. You've got to go out and scrabble around a little bit to get what you want.

When caribou are numerous, you should of course hunt caribou. When the Eskimo or the Indian hunter, skillful person that he is, finds that caribou are unhappily in short supply, he will simply redouble his efforts, for he is hungry, and back home the wife and kids are hungry. The whole situation is serious. The same tendency is found in every hunting-gathering mammalian predator activity.

On the other hand, an octopus has a much purer view. It behaves more like the theory of games predicts. When lobsters become few, an octopus does not seek in a frenzy to find those few lobsters or put up with eating mere crayfish. Heaven forbid! Instead the octopus goes to sleep—a most intelligent thing to do. Every once in a while it wakes up and looks out. "Any more lobsters around?" No. Back to sleep it goes again.

Such control over impatience, anxiety, and hunger is very hard for us to understand. Our thought is based on our design: namely, we have to generate one hundred watts of power all the time. There is a basal metabolism, roughly one hundred watts, that we expend. If we don't keep the machine fueled, we're in irreversible danger. But the octopus has no such base load demand. Cold-blooded, he is willing to relax to the ambient temperature of the warm sea environment in which he lives, provided only that every once in a while he can scrape up a fraction of a watt, turn the retina and its ganglia on, open an eye. That isn't too hard to do. Once you look at it coolly, you realize that human behavior goes completely against the sound principles by which an organism would adaptively go hunting. Whenever it is hard to hunt, don't continue to hunt with greater frenzy over longer hours, as we all do. On the contrary, take it easy. When conditions are not good, there is not much use in hunting. We humans can't adopt that principle, though, because for a couple hundred million years conditions were generally good enough so that somehow or other it was worth paying to keep our subtle electronics going—even during sleep—in order to have an opportunity to hunt well.

How different evolutionary structures can be! We need a special view of the lives of other creatures. If you now carry this logic over to some distant world, then it gives scope to the issues we are talking about.

The discussions of SETI show in detailed operational terms that we have already begun to make a clear plan. Certainly, a great deal of hope emerges. Indeed, we have a remarkable new opportunity: We seem to be on the threshold of finding other planets. We are setting up apparatus dedicated to the purpose of finding planets, by spectroscopy, interferometry, and by astrometry with modern techniques. I very much hope that the entire scientific community will support and applaud this effort, because it seems to be one of the most important auxiliary searches that could be made. It is important even if we never have a chance of getting radio signals. It can give us something else to look at than just the single Sun-planet system to which we are so well adapted.

We can characterize attitudes toward SETI by involving the names of two philosophers. First is the Aristotelian view, which seems quite plain: Earth is the cosmic center; the heavens revolve around it, 1/R reaches infinity here, and here is the right place! In this view, of course, the whole outward-looking style is neither necessary nor desirable. Astronomers can hardly accept that view; at least they have not accepted it for several centuries now. They are not going to change, and I am pretty sure that most of the other sciences will follow in turn.

The second point of view I like to attribute to Copernicus. Everyone knows what that name implies, though I don't know if he actually said it anywhere: namely, that this green-and-blue Earth is not all that different from the planets and the Sun and all those other things that circle and shine in the sky. They are themselves earthy or gaseous or whatever, but they are physically real objects. Nowadays men have walked upon one such object and shown that it is not different in kind from the one we inhabit. Since we know Earth is also earthy, then it is clear to us that these are only relative categories. Circular, shining, and perpetual orbits was Aristotle's view; it was Copernicus who recognized that Earth was no less circular, shining, and perpetual. We take a very different view of the cosmos post-Copernicus. That has been the spirit of SETI. The radical Copernicanism of the very first efforts is still viable, though admittedly we have more judgment about where to look.

A curious situation has arisen under the power of intense modern instrumental specialization. Our specialized tools and their data have grown steadily. That is sharply reflected in the institutions of our universities, which are slow to change in the face of it. For example, many universities still have a botany department and a zoology department, but if you bring them any one of a number of microorganisms, they can't say which department ought to study it! Perhaps it doesn't make any difference. But our research structure is inherited from institutional decisions in Scottish and German universities made 100 to 150 years ago. Sooner or later this will change. For all real large-scale engineering activities, such as NASA has carried out so successfully, we know that this is not the way to do it. Such activities require mission teams and a mix of specialties. The universities are going to have to learn, and some have begun.

There is another narrowness of action, though not of intent, which characterizes university departments, and scientific publications and scientists in general: If it is too popular, it is somehow vulgar and wrong. You can't really speak to those people across the street. I live next to the chemists at MIT, but I never see them. I hardly know who they are, yet between physics and chemistry it is hard to know who should study what molecule. I myself am guilty. We form communities not based on the problems of science, but on quite other things. This is part of the general split between the intelligent informed member of the public and the scientist who speaks in narrow focus. But the great theoretical problems that I believe the world expects will somehow be solved by science, problems close to deep philosophical issues, are the very problems that find the least expertise, the least degree of organization, the least institutional support in the scientific institutions of America, or, indeed, the world.

Two of these, of course, are the great questions, "Are we alone?" and "How did life begin?" These questions are treated now in the elementary textbooks, because of the vigor of a few people over the last thirty years, but they are hardly treated anywhere else. The further you go away from the freshman student, the less likely you are to find a colleague interested in it. This is beginning to change. Five or ten years ago, the radio astronomers, just to name a group of people I know quite well, were pretty hard to talk to about SETI in any way. It wasn't so much that they would disagree; that's fine—they still do. But they laughed, and that was not very pleasant. Well, now at least they are only smiling; that is a kind of gain.

One cure for this ill, though a difficult one, is the pursuit of a scientific discourse on a more philosophical, more consciously aesthetic, better-illustrated style, one willing to grapple with large problems, even though only small solutions can at present be offered for them. I think that science requires this mode too. I expect to see an enlarging of the disciplines to form at last an interdisciplinary pool, aware of larger philosophical issues. We need not try to solve them or to prescribe their limits, but we must recognize their human importance, their intellectual existence as an increasing element within scientific thought. If that were the only positive result from the SETI investigation, I think it would still be judged by history to have proved extremely worthwhile.

# Appendix A

# Amateur Equipment for SETI

## by D. Kent Cullers and William R. Alschuler

As mentioned in the chapter on Amateur Participation, there are various requirements for radio equipment to do SETI. We can only summarize here and give suggestions for meeting them. Detailed treatment is beyond the scope of this book.

### ANTENNAS

We have suggested that you use standard TV dish receivers, most of which are three meters in diameter (they can be purchased slightly smaller and also larger, up to four meters—the larger the better). For any type of SETI-observation technique you will need to point the antenna to various altitudes. The dish will feel varying forces and may be distorted as a result. If the dish distortion is too large, the focal sharpness will be ruined, which can lead to spillover and a wasted signal, as well as a shift of observed frequency. To check for this problem, you can run pieces of string from rim to rim across various diameters and chords, and check that they all touch where they cross, and stay touching at every pointing direction. If they don't, you need to reinforce the dish, which you should do with non-conducting materials. Also check to see that the feed does not sag out of the focal spot and that the dish shape is smooth. A non-pie-slice dish will likely be best in the latter respect.

### ANTENNA MOUNTS

There are too many variations of mounts to catalog, but some common characteristics do exist. Most have a main vertical post (some are tripod supported). The majority have a main tilted axis, the polar axis, which is set to point parallel to Earth's north-south axis. There is also often a stand-off yoke, which carries the dish at a slight tilt (often limited in range) to the polar axis. This tilt is to allow for your local parallax in viewing satellites; that is, the perspective shift you observe from being north (or south) of the Equator, looking up at an angle at satellites that are orbiting directly above the Equator. The majority also have a limited east-west axis adjustment, to allow pickup of the various satellites which are spaced out in orbit along the celestial equator.

To use the dish to track celestial sources, you need to align the polar axis accurately to the pole, and make the yoke parallel to it. Thus you need to align the polar axis so that its plane is in the meridian (north-south) plane. You can start alignment using a compass, but will probably need to check at night using the stars, or have in hand a good map that shows deviation of magnetic from true north (or south).

After that, you need to ensure freedom of rotation around the polar axis, for as much time as you wish to track potential ETI sources. Likely this would be for up to two hours, which would require rotation from fifteen degrees east to fifteen degrees west of the meridian. To automate this tracking, you will need to add a motor to the polar axis that tracks at the sidereal rate. You also need an axle to motorize, which may require you to modify the yoke or polar axis. To look at sources at any altitude above your horizon, you will need to have the freedom of pointing the dish from the pole to below the equator. This may require the unbolting of various actuator and lever arms.

Before changing anything on a dish that is operating well as a TV receiver, be sure to inscribe marks on the mounting so that it can be easily returned to proper alignment for TV, between SETI sessions.

## ANTENNA FEED

You need to check that your feed, which collects the radio energy at the antenna focus, will work well at the frequencies you have chosen for your SETI work. Alternate feeds can be purchased on the market.

## THE RECEIVER

A major component of your system is the receiver. In years past, a receiver with the right frequency coverage would have been a major problem, since the only commercially available receivers monitored police, fire, and amateur FM, but were useless for general purpose VHF and UHF listening. Now you can buy, for just under $1,000, a general purpose scanner that covers the region from 25 MHz to 2.0 GHz, which includes the water hole frequencies around 1.5 GHz. It's sensitivity is good but not optimal. Consequently, you should buy a Low Noise Amplifier (LNA) to connect between the antenna and your receiver. This combination will give you sensitivities comparable with those of professional radio telescopes.

The noise power in a band of frequencies is proportional to the bandwidth and to a quantity called the system temperature. In many common electrical circuits system temperature is directly traceable to the system's physical temperature. If, for example, one measures the current in a resistor, one finds random fluctuations resulting from the thermal motions of individual electrons. These random motions are excited by thermal behavior of molecules inside the resistor. As the temperature of the resistor increases, the random motions get bigger and the random component of the current grows. Thus, the noise in any given frequency band increases proportionately.

There is no intrinsic physical reason, however, that all circuits should behave like resistors. Though all of them yield outputs with random fluctuations, the constant of proportionality between these and the average output may be much smaller than those of simple resistive devices. Certain transistor devices, notably Gallium Arsenide Field Effect Transistors (GaAs FETs), tend to decouple the current carriers from thermal effects. Operating at room temperature (300K), amplifiers using these transistors have noise temperatures of about 100 K, about a third of that in a room-temperature resistor.

If you use as LNA a GaAs FET amplifier between your antenna and a broadband-scanning receiver, you can get effective noise temperature of 100 K plus the antenna temperature (typically the latter is 50 K; if you add shielding you can get to 10 K or so). This is a system that an observatory operating in the early 1960s would have envied.

So, first find an antenna. The better built the antenna, well-soldered connectors, smooth edges, etc., the lower the system noise temperature and the higher the antenna gain. Attach the antenna to the LNA with as short a cable as possible. Long runs of coax add to the system noise. After the transistor amplifier has increased the strength of the incoming signal hundreds of times, connection of the rest of the system is much easier. This is because the amplified signal and noise from the LNA dominates later stages in the system. The added noise of a slightly detuned receiver or too

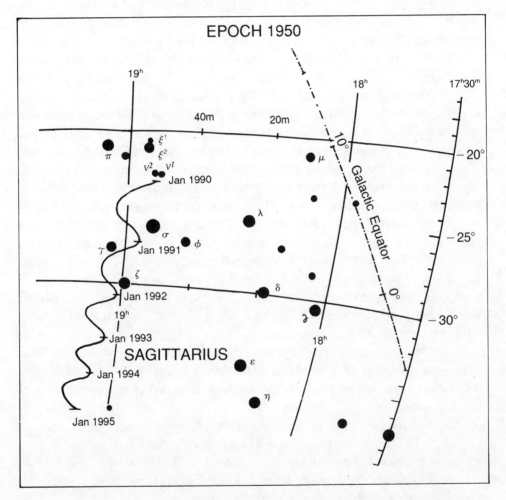

**Figure 1.** Voyager 2 spacecraft positions. This chart shows approximate positions for Voyager 2 over the next several years. Note the "teapot" of Sagittarius; the proximity of the galactic center. The Milky Way has been omitted for clarity. Voyager's zig-zag motion is the result of our changing viewpoint, as the Earth travels around the Sun. *(Diagram: W.R. Alschuler. Sky map after Norton's Sky Atlas, 17th edition.)*

long piece of coax is insignificant compared to the amplified roar already coming into the system front end.

We suggest for a receiver the IC-R7000, a general-coverage scanner. It is easy to use with a host of scanning modes. If you did nothing except scan frequencies throughout its range with your dish antenna pointed randomly at the sky, you would be exploring the unknown. Just listening to the receiver speaker, your search would, for most frequencies and directions in the sky, be thousands of times more sensitive than any performed to date.

You can listen in a selection of modes: AM, FM (good for diagnosing interference, but not if you want to look for drifting signals), and SSB (Single Sideband) are all available. This last mode makes our radio and TV carriers come literally whistling up out of the noise as you tune by. You can set the scan rate of the device to be fast or slow over specified limits. For an extra sixty dollars, you can even install a speech card in the scanner. Then, on every frequency where there is a signal, the receiver will stop and announce the frequency. Put a tape in a recorder, and the receiver will turn it on only when signal is present. Then, once a day, you can monitor a short tape of the "hits" for the last twenty-four hours. All this can be done with nothing beyond standard radio technology and an operator willing to spend some time. The receiver, without speech card, sells for slightly less than $1,000. It is made by Icom, a well-known manufacturer of amateur radio gear, and can be purchased in a ham radio store or mail order. For example, Henri Radio in Los Angeles, California, and Ham Radio Outlet in Oakland, California, carry this product.

The Low Noise Amplifier with connectors of your choice—I suggest BNC Type—can be purchased for about $100. When ordering, be sure to specify the frequency range of interest. As of this writing, LNAs can be purchased from, among others:

1. Bob's Electronic Service, 7605 Deland Avenue, Fort Pierce, FL 34591. Telephone: (407) 464-2118 (Robert Sickels).

2. Research Communications Limited, Unit 1, Aerodrome Industrial Complex, Aerodrome Road, Hawkinge Folkestone, Kent CT18, England. Telephone: 0303-893631.

The first store also carries equipment for the amateur astronomer, including complete radio telescopes with instructions and free advice.

DATA PROCESSING

If you want to do a Fourier Transform spectrum analysis on the output from the scanner speaker, you need a computer and a device to digitize the voltage so its value may be written into your memory. You must sample your data at more than twice the highest frequency coming out of your speaker. For the IC-R7000 in narrow, SSB mode, the sample rate must be at least 5,000 samples per second (5KHz). Whatever your computer, a digitizer board probably exists for it that can handle this modest sampling rate.

Since you have a computer anyway, you may want to computer-control your scanner instead of programming it from its front panel. This is no problem with the IC-R7000. Almost all of the receiver functions are addressable in ASCII.

Once data is in your computer, it can be analyzed to tell you how much power is at each frequency. Remember that the length of data you analyze determines the bandwidth of each frequency channel. The data should be run through a Fourier Transform program, available in many standard mathematical tool kits. An excellent choice is "Numerical Recipes," which comes in C, Fortran, and even Pascal versions. The output of the transform can be tested to see what sine wave amplitude is too high to be a random event in your system. Some experience will be required in order to set this threshold. Too low a threshold will keep you forever analyzing random noise and radio frequency interference. Too high a threshold will leave you in peace but may miss a weak signal.

TESTING THE RIG

You can check your rig's sensitivity, tracking ability, pointing accuracy, and your software (filtering algorithms) by looking at various targets. These include: the TV satellites themselves, perhaps with a signal attenuator; the moon, to pick up scattered signals of all types from Earth; bright pulsars; and the Pioneer and Voyager (and Galileo) spacecraft.

You may wish to start by observing standard celestial radio sources and to join SARA, an organization for amateur radio astronomers. That way you can get your feet seriously wet before trying the more demanding tasks of SETI.

VOYAGER II AS A SETI TEST TARGET

Once you get set up to look for ETI signals, as discussed above, it will be a good idea to practice on some known sources with characteristics similar to those you are hoping to detect. A demanding test would be to listen for the signals from one of our distant space probes on its way to interstellar space. We suggest trying to listen to Voyager II, currently about 30 astronomical units from us and moving away from the Sun at about 3½ A.U. per year. During Voyager's flyby the gravity of Neptune deflected it east and downward. Voyager II is now traveling systematically about 0.2 degrees eastward and about 2.0 degrees southward (below the plane of the ecliptic) per year. However, its apparent position against the background stars is also affected by Earth's orbital motion. This causes the spacecraft to undergo a back-and-forth, east-west cycle of about plus or minus 2 degrees at its current distance (much like the apparent motion of Neptune). This effect will diminish as Voyager's distance increases.

This means that, with a small radio telescope with an angular resolution of 10 degrees or so, you can expect to find Voyager II in your beam if you aim at: Right Ascension = 19 hours, and Declination = −30.0 degrees, for several years from now (1990). This position is just below the ecliptic, in the constellation Sagittarius, near the star nu Sagittari. This position is given for epoch 1950 star charts and will change slowly over the next several years, as described above. (Our thanks to Dr. Don Gray of the navigation team at JPL for supplying the position of Voyager II.)

You may have to modify your receiver slightly to pick out Voyager's signal, which is at about 8 GHz, well above the water hole frequencies. It is only broadcasting about 20 watts, so the signal will be quite weak. Good hunting!

**Figure 2.** Close-up of the Milky Way, in the direction of the galactic center. Our galaxy's central bulge of stars lies above and below the dark lanes of dust that obscure the center. The tail of Scorpius, including the bright star Antares (imbedded in nebulosity) and a globular star cluster, M 4 (just above and right of Antares), lie above the center. *(Photo: Courtesy American Museum of Natural History.)*

# Appendix B
# STARS NEARER THAN 5 PARSECS

| No. | Name | RA | Decl (1950) | Proper motion | Parallax | Distance light years | Visual apparent magnitude and spectral type |
|---|---|---|---|---|---|---|---|
| 1 | Sun | | | | | | − 26.8 G2 |
| 2 | α Centauri | 14$^h$36$^m$2 | − 60°38′ | 3″.68 | 0″.760 | 4.3 | 0.1 G2 |
| 3 | Barnard's star | 16 55.4 | + 4 33 | 10.31 | .552 | 5.9 | 9.5 M5 |
| 4 | Wolf 359 | 10 54.1 | + 7 19 | 4.71 | .431 | 7.6 | 13.5 M8e |
| 5 | BD +36°2147 | 11 00.6 | +36 18 | 4.78 | .402 | 8.1 | 7.5 M2 |
| 6 | Sirius | 6 42.9 | − 16 39 | 1.33 | .377 | 8.6 | − 1.5 A1 |
| 7 | Luyten 726-8 | 1 36.4 | − 18 13 | 3.36 | .365 | 8.9 | 12.5 M6e |
| 8 | Ross 154 | 18 46.7 | − 23 53 | 0.72 | .345 | 9.4 | 10.6 M5e |
| 9 | Ross 248 | 23 39.4 | +43 55 | 1.58 | .317 | 10.3 | 12.2 M6e |
| 10 | ε Eridani | 3 30.6 | − 9 38 | 0.98 | .305 | 10.7 | 3.7 K2 |
| 11 | Luyten 789-6 | 22 35.7 | − 15 36 | 3.26 | .302 | 10.8 | 12.2 M6 |
| 12 | Ross 128 | 11 45.1 | + 1 06 | 1.37 | .301 | 10.8 | 11.1 M5 |
| 13 | 61 Cygni | 21 04.7 | +38 30 | 5.22 | .292 | 11.2 | 5.2 K5 |
| 14 | ε Indi | 21 59.6 | − 57 00 | 4.69 | .291 | 11.2 | 4.7 K5 |
| 15 | Procyon | 7 36.7 | + 5 21 | 1.25 | .287 | 11.4 | 0.3 F5 |
| 16 | Σ 2398 | 18 42.2 | +59 33 | 2.28 | .284 | 11.5 | 8.9 M4 |
| 17 | BD + 43°44 | 0 15.5 | +43 44 | 2.89 | .282 | 11.6 | 8.1 M1 |
| 18 | CD − 36°15693 | 23 02.6 | − 36 09 | 6.90 | .279 | 11.7 | 7.4 M2 |
| 19 | τ Ceti | 1 41.7 | − 16 12 | 1.92 | .273 | 11.9 | 3.5 G8 |
| 20 | BD +5°1668 | 7 24.7 | + 5 23 | 3.73 | .266 | 12.2 | 9.8 M4 |
| 21 | CD − 39°14192 | 21 14.3 | − 39 04 | 3.46 | .260 | 12.5 | 6.7 M1 |
| 22 | Kapteyn's star | 5 09.7 | − 45 00 | 8.89 | .256 | 12.7 | 8.8 M0 |
| 23 | Krüger 60 | 22 26.3 | +57 27 | 0.86 | .254 | 12.8 | 9.7 M4 |
| 24 | Ross 614 | 6 26.8 | − 2 46 | 0.99 | .249 | 13.1 | 11.3 M5e |
| 25 | BD − 12°4523 | 16 27.5 | − 12 32 | 1.18 | .249 | 13.1 | 10.0 M5 |
| 26 | van Maanen's star | 0 46.5 | + 5 09 | 2.95 | .234 | 13.9 | 12.4 DG |
| 27 | Wolf 424 | 12 30.9 | + 9 18 | 1.75 | .229 | 14.2 | 12.6 M6e |
| 28 | G158-27 | 0 04.2 | − 7 48 | 2.06 | .226 | 14.4 | 13.8 m |
| 29 | CD − 37°15492 | 0 02.5 | − 37 36 | 6.08 | .225 | 14.5 | 8.6 M3 |
| 30 | BD + 50°1725 | 10 08.3 | +49 42 | 1.45 | .217 | 15.0 | 6.6 K7 |
| 31 | CD − 46°11540 | 17 24.9 | − 46 51 | 1.13 | .216 | 15.1 | 9.4 M4 |
| 32 | CD − 49°13515 | 21 30.2 | − 49 13 | .81 | .214 | 15.2 | 8.7 M3 |

| No. | Name | RA | Decl | Proper motion | Parallax | Distance light years | Visual apparent magnitude and spectral type |
|-----|------|-----|------|--------------|----------|---------------------|---------------------------------------------|
|     |      | (1950) |   |              |          |                     |                                             |
| 33 | CD − 44°11909 | 17 33.5 | − 44 17 | 1.16 | .213 | 15.3 | 11.2 M5 |
| 34 | Luyten 1159-16 | 1 57.4 | + 12 51 | 2.08 | .212 | 15.4 | 12.3 M8 |
| 35 | BD + 15°2620 | 13 43.2 | + 15 10 | 2.30 | .208 | 15.7 | 8.5 M2 |
| 36 | BD + 68°946 | 17 36.7 | + 68 23 | 1.33 | .207 | 15.7 | 9.1 M3.5 |
| 37 | L145-141 | 11 43.0 | − 64 33 | 2.68 | .206 | 15.8 | 11.4 |
| 38 | BD − 15°6290 | 22 50.6 | − 14 31 | 1.16 | .206 | 15.8 | 10.2 M5 |
| 39 | 40 Eridani | 4 13.0 | − 7 44 | 4.08 | .205 | 15.9 | 4.4 K0 |
| 40 | BD + 20°2465 | 10 16.9 | + 20 07 | 0.49 | .202 | 16.1 | 9.4 M4.5 |
| 41 | Altair | 19 48.3 | + 8 44 | 0.66 | .196 | 16.6 | 0.8 A7 |
| 42 | 70 Ophiuchi | 18 02.9 | + 2 31 | 1.13 | .195 | 16.7 | 4.2 K1 |
| 43 | AC + 79°3888 | 11 44.6 | + 78 58 | 0.89 | .194 | 16.8 | 11.0 M4 |
| 44 | BD + 43°4305 | 22 44.7 | + 44 05 | 0.83 | .193 | 16.9 | 10.1 M5e |
| 45 | Stein 2051 | 4 26.8 | + 58 53 | 2.37 | .192 | 17.0 | 11.1 M5 |

(Source: Rowan Robinson, Michael. *The Cosmological Distance Ladder,* New York: 1985. W.H. Freeman & Co.)

# GLOSSARY

**BANDWIDTH:** Any signal is broadcast over a certain more-or-less restricted range of frequencies, and any receiver is sensitive over a range of frequencies. These are called the transmitter and receiver bandwidths.

**BIOASTRONOMY:** The study of the signs of and conditions for evolution, and the possible structure of life on planets other than Earth.

**BLACK HOLE:** The core of a massive star that has gone supernova and whose collapse has continued past the neutron star stage to densities so extreme that space forms a "pocket" around it. Matter and energy are drawn into it and cannot escape.

**CHIRP:** A signal frequency is said to chirp (like a bird call) if it shifts up and down cyclically. This can occur if either the source or observer is in orbit or rotates. Thus the Earth's daily rotation will create a fairly rapid chirp of any received ET signals.

**DFT:** A Digital Fourier Transform chip is a semiconductor-circuit chip specially designed to rapidly split a signal into its various frequency components using the mathe-matical technique called a digital fourier transform.

**DOPPLER SHIFT:** The observed frequency of any signal is affected by the relative motion of observer and signal source. The greater the velocity, the greater the effect. For velocities of approach, the frequency increases; for velocities of recession, it decreases.

**ELECTROMAGNETIC SPECTRUM:** The array of gamma and x rays, ultraviolet, visible light, infrared, and radio waves spread out in order of wavelength and energy. All are electromagnetic particle-waves (photons) that propagate through space.

**KELVIN(S):** A temperature scale invented by Lord Kelvin, in which absolute zero (−273 degrees C.) is the base. On this scale, the liquid range of water is 273°K to 373°K.

**MAGIC FREQUENCIES:** The most abundant atom in the Universe, hydrogen, and the most important water-related molecular fragment, OH, both have signature radio line frequencies, under interstellar conditions, near 1,500 MHz. They seem like natural choices to broadcast near, if you want to catch the attention of ETs similar to us. Thus these are "magic" frequencies, and the frequency band between them is the "water hole."

**MULTICHANNEL SPECTROMETER:** A receiver-signal analyzer with the ability to collect signals in many different frequency ranges (usually narrow and side by side) simultaneously.

**NEUTRON STARS:** The leftover cores of massive stars that have gone supernova. They are extremely dense and often emit "lighthouse" beacons of radiation that we see as pulsars.

**NUCLEAR WINTER:** The rapid and prolonged cooling of Earth's climate, predicted as an unavoidable effect of even a limited nuclear conflict.

**NURSERY WORLD:** A planet ripe for life, undisturbed by external conditions.

**PLANETESIMALS:** Small agglomerations of rock in, or left over from, a forming planetary system, which may collect to form planets.

**RADIO ASTRONOMY:** The observation of celestial objects at radio wavelengths, using radio antennas and receivers.

**RADIO LINE:** Atoms and molecules have discrete internal-energy states unique to each type. A jump between states liberates (or absorbs) a photon of one frequency. For example, a cool cloud of hydrogen atoms emits many photons of 1420 MHz. If enough atoms emit similar photons, they show up as a spike, or (radio) line, against the general background static.

**STELLAR WIND:** The moderately high-energy stream of subatomic particles emitted steadily by the Sun and, perhaps, most stars.

**SUPERNOVA:** The death of a massive star through violent explosion, the outburst of which rivals a galaxy in brightness. In the explosion, heavy elements are created and scattered into interstellar space.

**XENOLOGY:** The study of alien life forms.

# SUGGESTED READING

## INTRODUCTION

COOKE, Donald A. *The Life and Death of Stars.* New York: Crown, 1985.

FERRIS, Timothy. *Coming of Age in the Milky Way.* New York: Morrow, 1988.

MCDONOUGH, Thomas R. *The Search for Extraterrestrial Intelligence.* New York: Wiley, 1987.

PREISS, Byron and Fraknoi, Andrew, eds. *The Universe.* New York: Bantam Books, 1987.

## CHAPTER ONE

BATESON, Gregory. *Mind and Nature.* New York: Dutton, 1979.

GARDNER, Howard. *Frames of Mind.* New York: Basic Books, 1983.

GRIFFEN, Donald. *The Question of Animal Awareness.* Los Altos, Cal.: Kauffmann, 1976.

WALKER, Stephen. *Animal Thought.* London: Routledge and Kegan Paul, 1983.

## CHAPTER TWO

GOLDSMITH, Donald. *The Quest for Extraterrestrial Life.* Mill Valley, Cal.: University Science Books, 1980.

SHKLOVSKII I. S., and Sagan, Carl. *Intelligent Life in the Universe.* New York: Dell, 1968.

*Time-Life* editors. *Life Search.* Alexandria, Vir.: Time-Life Books, 1989.

## CHAPTER THREE

BILLINGHAM, J., ed. *Life in the Universe.* Cambridge, Mass.: MIT Press, 1981.

MARX, G., ed. *Bioastronomy—The Next Steps.* Dordrecht, Neth.: Kluwer, 1988.

PAPAGIANNIS, M. D., ed. *Strategies for the Search for Life in the Universe.* Dordrecht, Neth.: Reidel, 1980.

## CHAPTER FOUR

BRAMS, Steven J. *Superior Beings.* New York: Springer-Verlag, 1983.

FINNEY, Ben and Jones, Eric. *Interstellar Migration and the Human Experience.* Berkeley, Cal.: University of California Press, 1985.

GOLDSMITH, D. and Owen, T. *The Search for Life in the Universe.* Menlo Park, Cal.: Benjamin Cummings Publishing, 1980.

MACVEY, John. *Interstellar Travel.* New York: Avon, 1977.

POYNTER, Margaret and Klein, Michael J. *Cosmic Quest: Searching for Intelligent Life Among the Stars.* New York: Antheneum Press, 1984.

TREFIL, J. S. and Rood, R. T. *Are We Alone?* New York: Scribner, 1981.

CHAPTER FIVE

DICK, Steven J. *Plurality of Worlds.* Cambridge: Cambridge University Press, 1982.

PAPAGIANNIS, M. D. *The Search for Extraterrestrial Life: Recent Developments.* Dordrecht, Neth.: Reidel, 1985.

PAPAGIANNIS, M. D. "Recent progress and future plans on the search for extraterrestrial intelligence." *Nature*, Vol. 318, (1985).

SCHENKEL, Peter. *ETI: A Challenge for Change.* New York: Vantage, 1988.

CHAPTER SIX

DRAKE, F. and Helou, G. "The Optimum Frequencies for Interstellar Communications as Influenced by Minimum Bandwidths." *NAIC Report 76.* (1978).

HOROWITZ, P. "Search for Ultra Narrowband Signals of Extraterrestrial Origin." *Science*, Vol. 201 (1978), 733–735.

OLIVER. B. "The Search for Extraterrestrial Intelligence." *NASA SP-419*, 63–73.

SULLIVAN, W. T., III. "Will the next supernova in our Galaxy be discovered with a radio telescope?" *Publications of the Astronomical Society of the Pacific*, Vol. 94 (1982), 901–904.

WOOLEY, R., Epps, E. A., Penston, M. J., and Pocock, S. B. "Catalogue of stars within twenty-five parsecs of the sun." *Royal Observatory Annual*, No. 5 (1970).

CHAPTER SEVEN

FASAN, Ernst. *Relations with Alien Intelligences.* Berlin: Berlin-Verlag, 1970.

HALEY, Andrew G. "Metalaw." *Space Law and Government.* New York: Appleton-Century Crofts, 1963.

MICHAUD, Michael A. G. "Extraterrestrial Politics." *Cosmic Search*, Vol. 1, No. 3 (Summer 1979), 11–14.

—— "Interstellar Negotiation." *Foreign Service Journal*, Vol. 49, No. 12 (December 1972), 10–14, 29–30.

SULLIVAN, W. T., Brown, S. and Wetherill, C. "Eavesdropping: The Radio Signature of the Earth." *Science*, Vol. 199 (27 January 1978), 377–387.

CHAPTER EIGHT

BILLINGS, Linda. "Is Anybody Listening?" *Final Frontier,* Vol. 1, no. 2 (June 1988).

CAMERON, A. G. W., ed. *Interstellar Communication.* New York: Benjamin, 1963.

COCCONI, G. and Morrison, P. "Searching for Interstellar Communications." *Nature*, Volume 184 (1959), 844.

COUSINS, Norman. "Why Man Explores." *NASA EP-125* (1976).

DRAKE, Frank, Wolfe, John, and Seeger, Charles, eds.: *SETI Science Working Group Report.* NASA Technical Paper 2244, 1983.

FIELD, G., ed.: *Astronomy and Astrophysics for the 1990's. Volume I: Report of the Astronomy Survey Committee.* Washington: National Academy Press, 1982.

GINDILIS, L. M., Dubinskiy, B. A., and Rudnitskiy, G. M., "SETI Investigations in the U.S.S.R." Report Presented at the XXXIX Congress of the International Astronautical Federation, Bangalore, India, 1988.

GREENSTEIN, J. S., ed.: *Astronomy and Astrophysics for the 1970's.* Washington: National Academy Press, 1972.

MORRISON, Philip, Billingham, John, and Wolfe, John, eds. *The Search for Extraterrestrial Intelligence.* New York: Dover, 1979.

SAGAN, Carl, ed.: *Communication with Extraterrestrial Intelligence.* Cambridge, Mass.: MIT Press, 1973.

CHAPTER NINE

BALLARD, J. *Handbook for Star Trackers.* Cambridge, Mass.: Sky Publishing, 1989.
CULLERS, D. K., Linscott, I. R., and Oliver, B. M. "Signal Processing in SETI." Communications of the ACM, (ACM/IEEE-CS Joint Issue) Vol. 28, No. 11 (November 1985).
TAUB, Herbert and Schilling, Donald L. *Principles of Communications Systems.* New York: McGraw Hill, 1971.

# RELEVANT ORGANIZATIONS

*The Planetary Society*
P.O. Box 91687
Pasadena, CA 91109
(The premier supporter of SETI research outside of government.)

*The Astronomical Society of the Pacific*
1290 24th Avenue
San Francisco, CA 94112
(A leading association of professionals and amateurs with interest in SETI.)

*Society of Amateur Radio Astronomers (SARA)*
P.O. Box 250462
Montgomery, AL 36105
Membership is $20/yr for U.S., $21 for Canada, $28 for other foreign countries.
(The major amateur organization in radio astronomy.)

*International Astronomical Union (IAU) —Commission 51 on Bioastronomy*
c/o Prof. Michael Papagiannis, secretary, Dept. of Astronomy, Boston University, Boston, MA 02215
(The IAU is the largest international professional umbrella organization in astronomy.)

The National Aeronautics and Space Administration (NASA) Office of Public Information.
NASA Headquarters
Code L
Washington DC 20546

You must know the material you want when calling, but there is no charge for U.S. citizens.

# CONTRIBUTORS

BYRON PREISS is editor of *The Planets, The Universe*, and *The Dinosaurs*. He is also co-editor, with William R. Alschuler, of *The Microverse*. He has collaborated on books and computer software with Arthur C. Clarke and Ray Bradbury. His project, *The Words of Gandhi*, won a 1985 Grammy Award, and his monograph, *The Art of Leo and Diane Dillon*, was a Hugo Award nominee.

BEN BOVA, besides being the author of more than seventy books of science and science fiction, has six times won the Hugo Award for best editor for his work at *Analog* and *Omni* magazines. He has also been a consultant to Woody Allen, Gene Roddenberry, and George Lucas on film and television projects.

WILLIAM R. ALSCHULER is the founder and principal of Future Museums, a museum-consulting firm providing program concept through final design for museums and exhibits with a science or technology content. Alschuler has a Ph.D. in astronomy from the University of California at Santa Cruz, and extensive university teaching experience in the sciences and energy conservation. He is currently a member of the physics faculty at FIT/SUNY. His first book for Byron Preiss Visual Publications, *The Microverse*, was published in the fall of 1989.

HOWARD ZIMMERMAN entered the publishing field in 1976 as the editor of *Starlog* magazine, which covers the fields of science fiction and fantasy. He helped create and edited *Future* magazine, an Omni-esque publication that preceded Omni. He has also written and edited books on cinematic robots, spaceships, and aliens. He joined forces with Byron Preiss in 1987 and has since edited science fact and science fiction books for adults and for young readers, including two books on dinosaurs.

ISAAC ASIMOV's total of published books has passed four hundred volumes. Although his Ph.D. is in chemistry, his popular nonfiction has covered almost every imaginable subject, from quantum physics to the Bible. He is a winner of the Hugo and Nebula awards for science fiction. Among his novels are the Foundation series (winner of a special Hugo for Best Science Fiction Series of all time) and the Robot series. He is also the father of the Three Laws of Robotics.

GREGORY BENFORD writes for the Encyclopedia Britannica in the areas of relativistic plasma physics and astrophysics. His highly regarded fiction work includes the novel *Timescape*, which won four separate awards. He was nominated most recently for a Nebula for *Great Sky River*. He is a member of the physics faculty at the University of California at Irvine.

LINDA BILLINGS is a Washington-based aerospace writer. She was the senior editor for space at *Air & Space/Smithsonian* magazine, and founding editor of a newsletter on space commercialization, *Space Business News*. She has also worked on the staff of the National Commission on Space (the Paine Commission) and the U.S. International Space Year Association. She is a member of the board of directors of Women in Aerospace and co-chair of the American Institute of Aeronautics and Astronautics National Capital Section's public-policy committee. She has a long-standing interest in the history and politics of the space program.

STUART BOWYER received his Ph.D. in physics from Catholic University of America in 1965, with a thesis in x-ray astronomy. In 1968 Dr. Bowyer joined the astronomy department at the University of California at Berkeley. At Berkeley he has developed a group that is primarily involved with x-ray, extreme ultraviolet, and far ultraviolet astronomy, and related topics in high-energy astrophysics, including a number of projects launched on rockets and on orbiting spacecraft. Dr. Bowyer is a member of the

American Astronomical Society, the International Astronomical Union, the Astronomical Society of the Pacific, the American Geophysical Union, and the Optical Society of America.

DAVID BRIN's second novel, *Startide Rising*, won both the Hugo and Nebula awards. He was awarded another Hugo (his third altogether) for *The Uplift War*, set in the same universe. When he is not writing, he is a university teacher of physics.

BRUCE CAMPBELL began his astronomical career as a graduate student at the University of Toronto. In 1979 he became the first resident astronomer at the Canada-France-Hawaii Telescope in Hawaii. After a sabbatical at the Meudon Observatory in Paris, he moved with his wife and three children to Victoria, Canada. There he pursued his research at the Dominion Astrophysical Observatory, and then joined the department of physics and astronomy at the University of Victoria, where he now heads the High-Precision Stellar Velocities Group. He is best known for the development of a high-accuracy method for measuring velocities of stars. In 1980 he and his colleagues began a project to detect extrasolar planets with this technique, and today they are widely acknowledged as the leaders in this field. In 1989 they received the Muhlmann Prize of the Astronomical Society of the Pacific for this pioneering work.

ARTHUR C. CLARKE's achievements and honors are both scientific and literary in nature. Chief among them are his fellowship in the Royal Astronomical Society, the Hugo, Nebula, and John W. Campbell awards, the 1982 Marconi International Fellowship, a Fellowship of King's College, London, and an Oscar nomination (with Stanley Kubrick) for *2001: A Space Odyssey*. In 1979 the president of Sri Lanka nominated him chancellor of the University of Moratuwa, which is also the site of the Arthur C. Clarke Centre for Modern Technologies. His almost sixty books have sold thirty million copies in twenty languages, and he is credited as the inventor of the concept of the communications satellite.

HAL CLEMENT, a native of Massachusetts, holds degrees in astronomy and chemistry, and taught these and other sciences at the high school level for forty years until his retirement in 1987. He has been writing even longer, and is best known for producing so-called "hard" science fiction, in which the problems faced by the characters and the methods used in their solution both fit as closely as possible to currently understood science. His best-known novel is the science fiction classic *Mission of Gravity*. He is a retired air force reserve colonel, having flown thirty-five combat missions as pilot and copilot in Liberator bombers during World War II.

D. KENT CULLERS is the signal-detection team leader in NASA's search for extraterrestrial intelligence. He received his doctorate in physics from the University of California, Berkeley, in 1980, and came to NASA immediately thereafter. During the past ten years he has investigated ways to detect weak signals in a background of Gaussian noise and, in collaboration with the signal-detection teams, developed highly efficient algorithms for SETI. He is an Extraclass Amateur Radio Operator, and in a fundamental sense, SETI is an extension of his lifelong quest for distant signals. Dr. Cullers is married with two children. His hobbies include radio, playing the guitar, chess, and a great deal of science fiction reading.

ROBERT DIXON received his Ph.D. in electrical engineering in 1968 from Ohio State University. He is presently assistant director of the Radio Observatory at Ohio State. He was a faculty fellow on project Cyclops (SETI systems design project) and has published numerous papers on faint celestial radio sources.

FRANK DRAKE is presently a professor of astronomy and astrophysics and acting associate vice chancellor for University Advancement at the University of California, Santa Cruz. He received a Ph.D. in astronomy from Harvard University. He was affiliated for many years with the National Radio Astronomy Observatory and, later, Cornell University. At Cornell he was the director of the Arecibo Ionospheric Observatory and for ten years the director of the National Astronomy and Ionosphere Center, which operated the world's largest radio-reflector telescope at Arecibo, Puerto Rico. In 1960 he conducted the first modern search for extraterrestrial intelligent radio signals, Project Ozma, which was carried out at Green Bank, West Virginia. He is presently chairman of the U.S. National Committee for the International Astronomical Union, the chairman of the board on Physics and Astronomy of the National Research Council, the president of the Astronomical Society of the Pacific, and the president of the SETI Institute, a nonprofit corporation that carries out research related to extraterrestrial life with support from NASA.

PAUL HOROWITZ received his Ph.D. in physics at Harvard University and is now professor of physics there. He has been senior research associate at NASA-Ames, and jointly developed SETI signal-processing software with that team. He is a world leader in electronics design, and his research interests span from cellular motility through pulsars to SETI.

MICHAEL J. KLEIN received his Ph.D. in radio astronomy at University of Michigan in 1968. His dissertation was based on extensive observations of the planets and their surfaces. He jointed JPL in 1969 as a scientist in the Space Sciences Division. He became supervisor of the Radio Astronomy Group in the Planetary Atmospheres Section in 1974. Dr. Klein has more than twenty years' experience in radio-astronomy research with special emphasis on the development of observational techniques and the application of microwave and submillimeter radio astronomical experiments in the study of solar system objects. He serves as program manager for SETI at JPL, and is currently serving on the Arecibo Science Advisory Committee for the National Astronomy and Ionospheric Center (Cornell University).

THOMAS R. MCDONOUGH is a lecturer in engineering at Cal Tech and an award-winning public speaker. He is also coordinator of the Search for Extraterrestrial Intelligence (SETI) program at the Planetary Society, the 100,000-member international organization founded by Carl Sagan and colleagues. The *Los Angeles Times* named him "one of the rising stars to watch" in 1989. He is the author of the nonfiction books *The Search for Extraterrestrial Intelligence* and *Space: The Next 25 Years,* and of the science-fiction novel, *The Architects of Hyperspace.*

MICHAEL A.G. MICHAUD is director of the Office of Advanced Technology at the U.S.

Department of State, where he is responsible for the foreign-policy aspects of space activities and other advanced technology issues. He is the author of eighty-five published works, sixty-two of them on space or extraterrestrial intelligence. His publications include the 1986 book, *Reaching for the High Frontier,* a history and analysis of the modern American pro-space movement. He is a member of the International Institute of Space law, the Institute of Aeronautics and Astronautics, the American Astronautical Society, the Aviation/Space Writers Association, and several other space-related organizations. He holds a master's degree in political science from the University of California at Los Angeles.

PHILIP MORRISON took his Ph.D. in theoretical physics at Berkeley in 1940. For almost twenty years after the war he was on the physics faculty at Cornell. Since 1964 he has been at M.I.T., where he is now institute professor emeritus. Among his accomplishments are serving on the Manhattan Project with J. Robert Oppenheimer and taking part in the Trinity test, the first atomic explosion. He believes that physicists owe their fellow citizens two services: better understanding of physics, and independent comment on modern war and how to avoid it through peaceful means. He has written several books and made films along both paths. Most recently he and his wife, Phylis, wrote and appeared in a TV series on PBS called *The Ring of Truth,* which examined how science knows what it knows.

MICHAEL D. PAPAGIANNIS was born in Athens, graduated from the National Technical University of Athens, and received his Ph.D. in physics and astronomy from Harvard University. He is a full professor and past chairman of astronomy, and the recipient of the Award for Excellence of Boston University. He is a member of the International Academy of Astronautics, the National Academy of Greece, a fellow of the AAS, a member and past chairman of the Space Science Panel of the NRC/NAS Associateship Program, and member of the Executive Committee of the Haystack Radio Observatory.

DIANA REISS is the director of research into dolphin communication and cognition at Marine World Africa USA in Vallejo, Cali-

fornia. She is also on the faculty of San Francisco State, where she teaches courses on human and animal communication.

THOMAS F. VAN HORNE became a volunteer with the Ohio State University SETI project in 1987. In 1988 he was appointed chief observer at the OSU Radio Observatory. He is in charge of the SETI project's data-analysis group and is co-author with Robert Dixon of a paper describing the project's activities presented at the 1988 Toronto SETI conference. Van Horne is a professional computer programmer and long time science-fiction fan.